Mikko Heimola
From Deprived to Revived

Religion and Reason

Founded by
Jacques Waardenburg

Edited by
Gustavo Benavides and Michael Stausberg

Volume 53

Mikko Heimola

From Deprived to Revived

Religious Revivals as Adaptive Systems

DE GRUYTER

ISBN 978-1-61451-454-1
e-ISBN 978-1-64151-377-3
ISSN 0080-0848

Library of Congress Cataloging-in-Publication Data
A CIP catalog record for this book has been applied for at the Library of Congress.

Bibliografische Information der Deutschen Nationalbibliothek
Die Deutsche Nationalbibliothek verzeichnet diese Publikation in der Deutschen Nationalbibliografie; detaillierte bibliografische Daten sind im Internet über http://dnb.dnb.de abrufbar.

Druck: Hubert & Co. GmbH & Co. KG, Göttingen
♾ Gedruckt auf säurefreiem Papier
Printed in Germany

www.degruyter.com

Summary

The study addresses theoretical questions pertaining to the complex dynamics of religious revivals and, as an empirical test case, explores the history of Christian revivals in early nineteenth century Finland. It builds upon earlier research on the evolution of cooperation and altruism, the psychology of social norms, and findings that suggest features of religion help solve social dilemmas in group behavior. It addresses questions on the identity of those participating in a revival, the experiential factors that spur this behavior, and how the emerging revivalist group organizes to address perceived threats and challenges to group cohesion.

The methodological approach in the study is twofold. First, it is argued that based on existing empirical studies and theoretical thought a synthetic picture of religious groups as cohesive norm units can be construed and extended into accounting for religious revivals. This part also includes a simulation experiment that tests the logical coherence of the ideas advanced. Second, the case of Finnish revivals is explored to demonstrate the plausibility of the aforementioned theoretical construct, and to demonstrate that it offers a more comprehensive explanation for their occurrence compared to previous scholarship. The analysis focuses on three local revivals: the southwestern Jumping Revival, the Ostrobothnian *Skörts*, and the Kuortane Revival of Jakob Wallenberg and Isak Vasu.

The main result of the study is that antecedent explanations offered for the Finnish revivals can be integrated under the present approach, reconciling apparent conflicts between them. This revivalism is interpreted as arising from shared experiences of societal upheaval and a hypocritical church establishment, which motivated ritualized and ecstatic behavior that aggregated into separatistic ritual groups. Once established, these groups were in a position to work on the socioeconomic deprivations of their members. The study corroborates the view of religious behavior as a potentially adaptive response to environmental challenges, but which requires selection acting on cultural norm groups to weed out nonfunctional groups.

Preface

This book is based on my academic dissertation, which I defended on August 15, 2012 at the University of Helsinki. As such, it is the result of three years' work – work that I enjoyed immensely. It is not often that one is given the possibility to fully immerse oneself in such fascinating fields as the evolution of cooperation or histories of revivalist movements and freely pursue the trains of thought that appear most promising and relevant to better understand these phenomena. This was made possible by the generous funding provided by the Kone foundation and several persons whose support was crucial to the process.

First to demand attention is the enthusiasm and support of my academic supervisor, Docent Ilkka Pyysiäinen, who responded positively when I first approached him with some half-baked ideas. As a scholar and scientist, one of his many laudable qualities is his willingness to embrace new ideas and judge them by their merits, even when they require reconsideration of past views. In my approach, I was very much guided by his example.

Equally eager to aid me has been my other supervisor, Docent Timo Honkela of Aalto University, whose research interests extend from the technological to the social-scientific. In our discussions he was able to simultaneously show a keen interest in the matter at hand and also offer wider perspectives to consider. He also generously invited me to participate in sessions and workshops with his technology students.

The third person who incessantly aided me in my work is Professor Tuula Sakaranaho, who has supervised my studies from my bachelor's thesis to the present day. I have much reason to thank her for the numerous occasions she has supported my strivings and ideas. I also thank Professor René Gothóni who, in addition to acting as the custos of my disputation, has despite our rather different methodological approaches offered very perceptive commentary on my seminar papers.

I also thank my two pre-examiners, Professor Richard Sosis and Dr. Kimmo Ketola, for their meticulous examination of my work. Commentaries from these two researchers have been especially valuable to me. Engaging with the work of Professor Sosis forms a central part of my dissertation, and his constructive feedback has helped to clear away some misunderstandings that would have hindered rather than advanced the discussions in this field. Dr. Ketola, researcher of the Church Research Institute and formerly of the Department of Comparative Religion, on the other hand, is well-versed in both the theoretical and the empirical substance of my work, and thus uniquely positioned to judge its flaws and merits. I also thank Professor Sosis for acting as my opponent in the public defense of my dissertation, and for his many kind words about it.

In addition to these people, during the process I have had the pleasure of discussing various aspects of my work with several well-informed persons, including but not limited to István Czachesz, Antti Salovaara, Tom Sjöblom, Atte Tahvanainen, Tuomas Tammisto, and Risto Uro. Docent Kimmo Vehkalahti is to be thanked for his generous help with the statistics. In addition to them, I thank my colleagues at the department, especially Riitta Latvio, Vesa Matteo Piludu, and Outi Pohjanheimo, with whom I have shared a study (and many discussions) during these years. The aid of amanuensis Riku Hämäläinen and department secretary Sonja Pakarinen of the Department of World Cultures has been considerable, as has been that of Kristiina Norlamo, administrative officer of the Faculty of Arts. The department has also been generous in giving me a desk to work at.

I am grateful to Professor Michael Stausberg and Professor Gustavo Benavides for accepting my work in the Religion and Reason series, and to Dr. Alissa Jones Nelson and Sabina Dabrowski of De Gruyter for their help and assistance during the publication process.

I also thank my parents Antti and Marketta, who have continued to support my professional questing despite its winding course, my lovely sisters Laura and Erika, and my numerous friends with whom I engage in activities that constantly widen my perspectives. But most important in keeping me from even thinking of this project outside the workplace have been my two wonderful sons, Aatos and Touko. Rather than being a burden to my careerism, they have considerably advanced it by enforcing clear boundaries to it. And truly, the last and most important person to remember is my 12-year companion, colleague, and wife, Minna, without whom none of this would be.

At my home in snowy Helsinki,
February 18, 2013
Mikko Heimola

Contents

1 Introduction

> It should be known that history, in matter of fact, is information about human social organization, which itself is identical with world civilization. It deals with such conditions affecting the nature of civilization as, for instance, savagery and sociability, group feelings, and the different ways by which one group of human beings achieves superiority over another. It deals with royal authority and the dynasties that result in this manner and with the various ranks that exist within them. Also with the different kinds of gainful occupations and ways of making a living, with the sciences and crafts that human beings pursue as part of their activities and efforts, and with all the other institutions that originate in civilization through its very nature. (Ibn Khaldûn [1377] 1967, 35)[1]

1.1 Subject and Plan of the Study

The subject of the present study is the complex dynamics of cooperative bonds exhibited by religious revivals. Even briefly put, it is affiliated to at least three separate areas of scientific or scholarly study: first, that of complex adaptive systems, an interdisciplinary field of inquiry that focuses on the systemic patterns that appear to recur on all levels of empirical reality, from behavior of physical matter to social organisation, whether of honeybees or of humans (Epstein and Axtell 1996; Sun 2006; Miller and Page 2007; Tautz 2009); second, the game-theoretic interpretation of the multidisciplinary puzzle of cooperation, prosociality, and group action, that has traditionally interested mostly economists (Olson [1965] 1971; Hardin 1968; Greif 1989, 1993; Ostrom 1990; Frank 1993), evolutionary biologists (Trivers 1971; Axelrod and Hamilton 1981; Sober and Wilson 1998; McElreath and Boyd 2007; Drea and Carter 2009), and social psychologists (Tajfel 1970; Dawes 1980; Orbell, van de Kragt, and Dawes 1988; Sally 1995; Yamagishi, Jin, and Kiyonari 1999), but has in recent years attracted increased attention also from other disciplines (Nesse 2001; Norenzayan and Shariff 2008; Pyysiäinen 2010a) and attempts towards grand synthesis (Hammerstein 2003;

1 'Abd-ar-Rahmân Abû Zayd bin Muhammad bin Muhammad bin Khaldûn (1332–1406) was an Islamic scholar, whose best-known work, *Muqaddimah* (1967; actually the introduction to his universal history), Dawood regards as "the earliest attempt made by any historian to discover a pattern in the changes that occur in man's political and social organization" and "an almost complete departure from traditional historiography, discarding conventional concepts and clichés and seeking, beyond the mere chronicle of events, an explanation – and hence a philosophy of – of history" (Dawood's introduction in Ibn Khaldûn 1967, ix). Modern theorists have also been appreciative of his work, most notably Turchin (2009, 195; Turchin and Nefedov 2009, 23–25). Due to the affinities his argument has with the present work, I have chosen introductory quotations from the *Muqaddimah* for my chapters.

Nowak 2006; Henrich and Henrich 2007); and third, the study of religion as a social, cultural, and behavioral phenomenon, where after what some view as the failure of secularisation theories (Norris and Inglehart 2004) an increased attention has focused on the ubiquity of religious behavior across space and time, and the role it has in the evolution and organization of human affairs (Wilson 2002; Atran and Norenzayan 2004; Alcorta and Sosis 2005; Norenzayan and Shariff 2008; Pyysiäinen 2009; Atran 2010).

I propose that by uniting the intellectual traditions above into a theoretical synthesis, we can get better answers to empirical questions about the role of religion in the organization of human affairs. To support my claim I present a case study of the Christian revivals that occurred in late eighteenth century and early nineteenth century Finland. Instead of a single large revival there were several small ones, although some eventually grew to have regional or even national significance. These revivals and the movements that sprang from them – Supplicationism (*rukoilevaisuus*), Laestadianism (*lestadiolaisuus*), the Awoken (*herännäisyys* or *körttiläisyys*) and their offshoot the Evangelical movement (*evankelinen liike*) – are rather well studied from perspectives of church history and social history. However, there does not appear to exist any widely accepted answer to the question of why the revivals happened and even those that have attempted to answer this question have run into difficulties in covering all the aspects, most notably that of the ecstatic phenomena that accompanied the revivals everywhere. Thus, in my analysis and review of relevant studies, my aim is intellectual integration: I attempt to demonstrate that in light of a new theoretical synthesis, most or all of the answers provided by previous scholars fall into place to form a unified view of this revival process.

The issues addressed by my study are of great importance to the comparative study of religions as an academic discipline. In addition to discussing traditional questions in the field such as reasons for religious change, it points to how the study of religion can contribute to solving what E. O. Wilson (1975, 382) called "the culminating mystery of all biology": how humans (who often cooperate even with strangers), out of all vertebrates, have been able to reverse what he terms the "downward trend" of cooperativity when descending from lower invertebrates (capable of coalescing into a single physical organism) to vertebrates (whose social life is mostly reserved to tending close kin). Almost all major evolutionary transitions from molecules to animal societies have required solving the problem of cooperation (West, el Mouden, and Gardner 2011, 235; also Michod 2007; Stearns 2007), hinting that religion may have had a unique role among humankind's cultural abilities in this reversal, leading many students of biological fields to consider functions of religiosity (e.g. in Bulbulia et al. 2008; Schloss and Murray 2009). But to produce a comprehensive account of these issues, contributions from students of culture and society are also needed to provide a civilizatio-historical perspective on the emergence of various forms of social organization over time and how these have contributed to human survival in changing ecological conditions (cf. Hietaniemi 2005, 15).

The plan of the study is as follows.[2] In the remainder of chapter one, I review the intellectual threads that make up what I consider the epistemological context of my study. These involve both system theory and evolutionary theory. The former serves both as a general philosophical grounding of my study and also more specifically as motivation for my methodological choice of social simulation. The latter – maybe the best-studied mechanism that drives complexity in real-world systems (Miller and Page 2007) – has become increasingly important in accounting for the push and pull between uniformity and variation that operates on multiple levels of human life, including religions of the individual and of society. These paradigmatic underpinnings allow me to describe how I perceive the link between the psychological and the social, which is crucial for the present study. I also attempt to demonstrate what my thesis is and is not, *vis-à-vis* the antecedent study of religion: the outward form may appear different, but some of the core ideas are very similiar to those advanced by Durkheim (1915), Weber (1930), and Rappaport (1979, 1999), while the contrary is true concerning some rational choice and economic accounts of religion. By these clarifications I hope to make my thinking more accessible to those who are less specialized in the particular subfields I employ.

In chapter two, I describe cooperation as a multi-disciplinary issue of great import, both scientifically and for human condition in general. I review the general mechanisms for evolution of cooperation proposed by evolutionary theorists and discuss whether these can explain the large-scale cooperation that characterizes human activity. It appears that to account for the variability of human behavior across societies, a theory of social norms is required, and one following

2 Parts of this work have appeared or have been presented elsewhere. A draft version of chapter 4 entitled "A Signaling Model of Religious Change" was presented at the XXth Quinquennial World Congress of the International Association for the History of Religions in Toronto, Canada on August 19, 2010. The simulation experiment in chapter 4.4 was presented in the workshop "Computational modeling in social and economic sciences: evaluation of models of complex phenomena" of the International Conference on Artificial Neural Networks in Espoo, Finland on June 14, 2011, and in the meeting of the Nordic network "Socio-Cognitive Perspectives on Early Judaism and Early Christianity" in Järvenpää, Finland on June 18, 2012. An article covering parts that deal with emotional phenomena (chapters 2.4, 3.2.3, and 6.2.2) is to appear in the journal *Method and Theory in the Study of Religion* under the title "Threats of Hellfire and Jumping for Heavenly Joy: Manifest Emotionalism as a Solution to the Problem of Machiavellian Hypocrites in Religious Revivals" in 2013.

the outline of Bicchieri (2006) will be provided in this chapter. As a special issue of relevance to the present study (considering the salience of ecstaticism in religious revivals), I also review the strategic role of emotions in communicating commitment.

In chapter three, I review the evidence concerning a connection between religiosity and prosocial behavior. Based on this, a more specific operational definition of these terms can be formulated, e. g., what aspects of religion are relevant and how the resultant prosociality appears to be bounded. Theoretically the mechanism by which this connection operates appears more complicated than suggested by some earlier analyses, and to move forward I provide an in-depth analysis of key problems, such as why religion is different from other ideologies, and how social norms and emotions fit into the picture.

In chapter four, I set out an original theoretical formulation on how the ideas and findings of the previous chapters can be employed to account for the phenomenon of religious revivalism. I review empirical studies on conditions in which religious activism appears to surge and theoretical constructs both psychological and sociological that have been put forth to account for this. Based on these, I propose a reformulation that incorporates both existing and novel theoretical insights to account for both the dramatic inception of revivals and their lingering quality (e. g., if religions are to be seen as responses to environmental conditions, why they prevail even after these change). The chapter also includes an original simulation experiment, which gives support to some of the ideas advanced.

In chapter five I present the Christian revivals that spread through Finland in the late eighteenth and early nineteenth centuries as a case study of the theoretical ideas formulated in the previous chapters. I introduce the specific historical conditions (demographic change, societal reforms, economic stagnation) that are of relevance to my thesis and analyze in depth the history and beliefs of the three revivals: the Jumping Revival of Satakunta (*hyppyherätys*), the Kuortane Revival led by Jakob Wallenberg and Isak Vasu (*vasulaisuus*), and the Ostrobothnian *Skört* Revival (*körttiläisyys*). I also briefly discuss the scholarly debates that have surrounded the study of Finnish revivals, especially concerning the apparent juxtaposition between sociohistorical and theological traditions, and on what grounds the numerous scattered revivals can be characterized as a unitary phenomenon.

In chapter six I take a more analytic look at how the formation of revivalist groups can be better understood in light of my theoretical apparatus. I review earlier historical research on what is known of their social composition, mostly in the light of charges laid on conventicle-goers in a series of mass court proceedings. Concentrating mostly on the three aforementioned revivals, I also discuss what the documents tell us of their norms and practices and how ecstatic phenomena were employed in their doctrine. The focus here is on how the revived defined their in-group *vis-à-vis* the surrounding populace and what kind of operational environment they were able to create.

Chapter seven contains a summary of findings and a reappraisal of sociohistorical explanations on the Finnish revivals put forth by antecedent scholars. I also address some specific predictions made in the theoretical chapters, and attempt to clarify the nature of the self-organizatory process in relation to competing theoretical views. In the end, I suggest some prospective avenues for future research that could not be properly addressed within the present work, most importantly the role of charismatic leadership, the revivals' connection to the enclosure process (*isojako*), and why the surviving revival movements have developed in divergent directions since their inception.

1.2 The Intellectual Context

1.2.1 The Perspective of Complex Adaptive Systems

The Greek origin of the word "system," *συστημα*, means a "whole compounded of several parts or members" (Liddell and Scott 1940), something demarcated from the non-system while possibly static. But when these parts are endowed with the property of being able to react to changes in their environment, it becomes a special case, an adaptive system (Latin *ad*, towards; *apto*, to fit, to adjust, or to prepare) able to dynamically react to stochastic pressures. More difficult is the notion of a complex adaptive system. According to Miller and Page (2007, 3, 9), to understand such a system on the most basic level, it is not enough to know the workings of its every component. Rather, these systems exhibit behavior that is characteristic of the system level only, not of the component level. This arises from the dependencies between the components. A complex adaptive system exhibits robustness to changes in its environment or parts and can continue functioning even after being severely diminished in size. But a more radical change, such as the removal of a property crucial to the interaction between its parts, can reduce it to a non-functional system.[3]

3 For example, a beehive, the standard example of a complex adaptive system, can take a number of disruptions, such as a rodent gnawing its way inside the hive (they kill it with their stings and coat it with propolis, a resin-like substance, to contain infections), and quickly replace workers – or even the queen – that perish in these functions (Tautz 2009, 146–151, 196). But should there be trouble in some of the decentralized processes that orchestrate the division

How then can one demarcate the system from its environment? Depending on the point of view, this question may appear either trivial or unanswerable (e.g., does the bee system include the hive structure?), but here analytically the most pressing problem pertains to human social systems only. In a global network, can one social system be separated from another, and does one social system constitute a part of the environment of the other? The work of Luhmann (2006; Seidl and Becker 2006) has addressed this issue by treating social systems as communication systems that self-reproduce the distinction between themselves and the non-system. Communication is the privileged operation that upholds the system by creating a network in which the agents' selected information is constantly being encoded, decoded, and re-encoded to produce understandings of each others' intentions. In this view, the limit of the system can be situated on interaction points where the recipient agent no longer understands or chooses to retransmit the other's information. This creates an *emic* distinction between the social system in question and surrounding social systems, although one that only applies on a specific level of shared meanings.[4]

Distributed processes make a system self-organize so that it appears as if some central authority were guiding its behavior, but in a complex adaptive system, no such authority is necessary.[5] Here we approach the important concept of

---

of labor in a beehive, such as a new virus or a pesticide interfering with the adult foragers' production of the pheromone ethyl oleate, whose level regulates the maturation of in-hive nurse bees to foragers (Leoncini et al. 2004), the hive can face collapse if every worker quickly matures past the nursing stage, which leaves the brood without care.

4 This idea accords well with the established view in social psychology that every person entertains within him- or herself multiple roles or identitites (religious identity, gender identity, job identity) that filter and modulate communications in situations that cue a specific identity. Conflict between identities can be resolved in multiple ways (Roccas and Brewer 2002), but for the present study it suffices to note: 1) that humans appear capable of perceiving meaningful demarcations, however fluid and interweaving, within the massive social system of human population, and 2) that in practice every time I make reference to the environment of a social system, it also includes other social systems with which the present system has numerous micro-level interactions.

5 The above example of bees organizing their division of labor is a prime example: If many foragers are lost – they double as defenders – the amount of ethyl oleate the nurses receive drops (a feedback loop; Miller and Page 2007, 50), causing more of them to develop quickly into foragers, which restores the balance (Leoncini et al. 2004). There is no identifiable decision to reorganize the hive made by anyone (certainly not by the queen). In human societies, self-organization is observable from the level of informal communities setting controls on members' behavior without any identifiable authority (Ellickson 1991), to distributed governance of common pool resources more effectively than a central authority can manage (Ostrom 1990), and to the level of world trade organizing into networks and subnetworks beyond individual or institutional control (Abu-Lughod 1989).

emergence: the appearance of phenomena on the system level that seem disconnected from their origins on the component level. This echoes the definition of complex adaptive systems above, but despite its intuitive appeal it has proven hard give an objective and precise definition of what counts as an emergent phenomenon. Miller and Page (2007, 41–42, 234) suggest that it can be identified by having "a statistical signature of its own," that we can more parsimoniously predict its appearance and behavior by observing it on the macro level rather than by making micro level measurements and deductions (e.g., predicting the behavior of a hurricane from barometric pressure and humidity). This can be said of much of our information concerning the natural world, which appears to us not as an infinite mass of particles interacting in complicated ways but as solid-looking wholes whose behavior is studied by a number of differentiated sciences.

It is a commonly expressed view that empirical reality can be arranged into hierarchical layers (Anderson 1972, 393; Pyysiäinen 2004b, xvii; Miller and Page 2007, 40–42) (e.g., that physical particles combine into organizational units of chemistry, and biochemical processes in the brain bring forth psychological processes of memory and personality). Each step upwards in the hierarchy introduces an increase in complexity, as novel concepts appear for which a satisfactory explanation cannot be offered using only the conceptual tools of the lower level (McIntyre 2007). This contrasts with strict and "greedy" reductionist arguments, which claim that every level could be simply described in terms of particle physics once we have sufficient knowledge of their workings (Dennett 1995, 80–83); that, for example, to explain the behavior of a closed religious community, it would be enough to describe the features of individual psychology that make us capable of having beliefs in deities, participating in rituals, and fulfilling the needs to feed, reproduce, etc. Such a description commits the fallacy of assuming that reductionism implies constructionism (Anderson 1972, 393). To be more complete, it would have to include a parallel approach where an attempt is made to track how these more basic behaviors combine to create a societal-level entity whose behavior is no simple extension of individual needs, but unfolds as a meaningful "history" in itself. As Malley (1995, 5) opined, it is at least partly from the neglect of this latter approach that reductionism has a bad name in the humanities.

This may appear to some as an argument for methodological holism, but even though its father Émile Durkheim is often taken to have been against psychological and reductionist explanations of social life,[6] his argument for this comes very close to being a complex systems argument ([1895] 1982, 129):

> By virtue of this principle, society is not the mere sum of individuals, but the system formed by their association represents a specific reality which has its own characteristics. [...] By aggregating together, by interpenetrating, by fusing together, individuals give birth to a being, psychical if you will, but one which constitutes a psychical individuality of a new kind. [...] The group thinks, feels and acts entirely differently from the way its members would if they were isolated. If therefore we begin by studying these members separately, we will understand nothing about what is taking place in the group.

However, in practice, complex social systems are often studied by agent-based models (computer simulations), in which macro-level "social facts" are expected to emerge from micro-level "psychological" interactions between the agents. Here it is often only the individual agent's behavior that has actual causal power – not any imposing social facts – which is in keeping with the methodologically individualistic position. Gilbert (1995, 127) considers this a shortcoming in the current simulation practice, as "people are routinely capable of detecting, reasoning about and acting on the macro-level properties (the emergent features) of the societies of which they form part" (emphasis removed). Incorporating such macro-to-micro feedback mechanisms into models would help bridge the methodological gap. Next, I introduce the modelling approach, with special emphasis on agent-based models.

1.2.2 The Modelling Approach

A model is a depiction of a phenomenon that is simplified in comparison to the original, while retaining the information relevant to solving the problem of interest.[7] This in itself is saying very little about what the actual model looks like, but

---

6 In demeaning psychological explanations ("every time a social phenomenon is directly explained by a psychological phenomenon, we may rest assured that the explanation is false"), Schmaus (2010, 104–106) considers Durkheim to have been mainly opposed to the faculty psychology of Comte and Spencer, which he saw as simplistic verbal substitutions of biological urges for social phenomena. It was the introspective, unscientific methodology of the time he saw as problematic, not emerging experimental psychology.

7 More formally, if the system of interest can be thought of as consisting of a set of states S, the aim is usually to find the transition function $F(S)$ that describes how the system state changes over time. In practice there are often too many states, and the transitions between them are very complex, so states that are (for the present purposes) equivalent are mapped into aggregate states s using a function $E(S)$. This smaller set of states is used to construe the transition

the various forms it can take are governed by a tradeoff between precision and flexibility (Miller and Page 2007, 78–80, 223–224). Mathematical methods are very precise, but they can be meaningfully applied only to a limited number of problems. Verbal descriptions can be applied to every imaginable phenomenon, but their precision and verifiability is often questionable. In between the two fall computational methods, which can take many forms, from numerical simulation of intractable mathematical formulas to imaginative constructions of possible worlds. Mäki (2009, 4–6) gives a more exact description of the modelling approach:

> Agent *A* uses object *M* (the model) as a representative of target system *R* for purpose *P*; addressing audience *E*; at least potentially prompting genuine issues of resemblance between *M* and *R* to arise; describing *M* and drawing inferences about *M* and *R* in terms of one or more model descriptions *D*; and applies commentary *C* to identify the above elements and to align them with one another.

Mäki makes three important points here: 1) the model must be presented in a way that addresses the expectations of an audience *E*, so as to promote agreement and revision of beliefs – otherwise it fails in representing anything meaningful; 2) the resemblance between *M* and *R* is about specific elements of the two (as set out in *D*), not about any arbitrary way they may or may not match; and 3) the model *M* does not represent anything by itself, as it cannot be a perfect, unproblematic image of reality. A commentary *C* is required to align it with *R*, *P* and *E*.

This account of modelling goes against "the perfect model model," which assumes natural reality can be modelled exactly and comprehensively, e. g., by interpreting models as isomorphisms between theories and reality (French and Ladyman 1999, 112–114). But there are no universal criteria for determining simiiliarity between a model and the modelled, nor its approximate truthness (Teller 2001, 399–406). Instead, "[t]he more explicitly we formulate our theories [...] the more wrong they become" (Turchin 2003, 7), since more and more of the complexity in nature must be left out. In theoretical biology, it is a common practice to employ a mathematical or computational model as a plausibility argument for

---

function governing the model $f(S)$. The modelling effort is succesful if $f(E(S')) = E(F((S'))$, that is, the model ends up in the same state regardless of whether we first map the actual system state S' into a model state and apply the model transition function, or if we let the system state S' develop into its successor state and then transform this into a model state. So it is the choice of equivalence classes (how the empirical objects are implemented as model constructs) and the transition function (how we formalize our hypotheses on what are the governing laws of nature) that is essential (see Miller and Page 2007, 36–40).

some mechanism or process that is possibly working in nature, not as an exhaustive explanation (Plutynski 2001, 227). This is because of the historicity and complexity of biological systems, something they certainly share with sociocultural systems. In this endeavor, the model serves not so much as a representation of reality but as an independent investigative tool (Plutynski 2001, 229) or an “epistemic artefact” that materializes a postulated relationship into an observable form (Knuuttila 2005, 1266–1268). It is this latter property that allows models to transcend the reasoning abilities humans have – even to move beyond “anthropocentric” methodology (Humphreys 2009, 616–617).

However, the epistemic value of a model must also depend on whether it validly addresses the real world issues a given agent, purpose, and audience suppose it does. Realism cannot be completely disregarded, even if we take the more pragmatic view of models argued above. To reconcile this with the fact that models are never completely realistic, Mäki (2009, 9) suggests a “functional decomposition approach” in validating a model: “Models are not candidates for truth as wholes, rather their privileged parts can be considered for truth.” But in distinguishing between the relevant parts of a model, its purpose *P* and intended audience *E* must be considered, as they dictate which aspects of the model must have resemblance to the target system in order for it to achieve representativeness. This means that there are (avowedly) things in the model that are idealizations or plain falsehoods. But this is certainly something they share with controlled laboratory experiments, be they psychological tests of sociocultural behavior (cf. Wiessner 2009) or studies of evolution in action with *E.coli* (Plutynski 2001, 230–233).

Recently, some authors have considered computer simulations to be comparable to the experimental method (with reduced realism but increased control) or situated in-between theorizing and experimentation (e.g. Winsberg 2003; Peck 2004). A counter argument to this rather radical suggestion would include noting that models do not investigate nature but rather the validity of our own thinking (Kokko 2007, 7). But it is especially the agent-based simulation that has philosophical novelty and that does not easily fit into pre-existing methodological categories (Humphreys 2009, 619, 625). Epstein and Axtell (1996, 1–4) consider these to offer a uniquely new way of studying complex social phenomena – growing artificial socities – as they allow us to aggregate our theories concerning different areas of behavior (e.g., economic and religious) into a unified whole and test the compatibility of micro- and macro-level theories of individual and collective behavior.

A typical agent-based model consists of three principal components (Epstein and Axtell 1996, 4–5):

- *Agents* model (human) individuals. Each of them has its own internal state (modelled by a number of variables such as gender, wealth, and social identity) that is subject to change in the course of simulation.
- *Environment* in which an agent engages in its behavior. At a minimum it consists of other agents, but it may also contain other meaningful points of interaction (such as resources to gather). Its topology need not be that of a two-dimensional map, but can also be a (social) network or something more abstract.
- *Rules* govern agents' interactions with the environment and other agents: how they form contacts, engage each other, or seek resources; and how their internal state changes. The environment can have its own rules, such as how long it takes for a harvested resource to renew.

In addition, the model needs an initial stage, after which the rules govern the evolution of the system over time. Usually the aim is to generate recognizable macro-level phenomena such as polities, exchange systems, or demographic oscillations. If this is achieved, the simulator has construed an existence proof for his theorem: given these preconditions, the postulated result can happen (Kokko 2007, 183; Miller and Page 2007, 86). Of course, for an elegant proof, the same simulation should not generate any contradictory or ill-defined phenomena (with respect to its privileged parts, cf. Mäki 2009, 9, quoted above).

A potential problem identified with simulations relates to the fact that they allow the modeller to implement his or her hypothetical scenarios in considerable detail (up to creating complex artificial worlds, cf. Epstein and Axtell 1996). This may not be so desirable in practice, as every additional variable involves making further (sometimes arbitrary) decisions as to how to properly formalize it. Even if we are after surprising, emergent phenomena, not being able to interpret the causal chains behind them leaves us with a model we don't understand in addition to the world we didn't understand (Kokko 2007, 10–11; McElreath and Boyd 2007, 10). Analytical models (specified as mathematical expressions) do not have these issues, but many real-life problems are mathematically intractable (e.g., those that involve networks). Thus simulations may often be the only alternative. Also, in contrast to purely mathematical methods, simulations allow inspection of not only equilibrium states but the dynamics and change to which the model is subject (Axtell 2000, 11–12, 15).

Another oft-mentioned criticism relates to the cognitive simplicity of the agents, although while spelling this out, Atran and Medin (2008, 152–159) consider the approach readily compatible with the cognitively more realistic views of cultural epidemiology (Sperber 1996) and situated cognition (Strauss and Quinn 1997). While in part methodologically motivated (what can be generated from trivial cognitions), this simplicity is dictated by computational demands. Sun and Naveh (2007, 124) employed the cognitively sophisticated CLARION architecture in a social simulation of survival strategies, but were forced to limit the number of agents to 30 – a very low number compared to the thousands required in some other simulations for interesting effects to emerge. Thus, while cognitive modelling is an important area in itself (see Sun 2006), for the time being, there also has to be room for large-scale simulations of simple agents.

In the modelling strategy I employ (in chapter 4.4) we can identify deductive, inductive, and also abductive logic at work (Sugden 2000, 19–25). Building and operating a model is mostly an exercise in deduction. In the model world, once the parts of the model are specified (regardless of their truth value), the outcomes are deduced from these according to the rules of formal logic. But then an inductive leap is required, when I claim that the observable outcomes (emergent phenomena), at least in their "privileged parts," bear some resemblance to what empirical studies tell us are real world phenomena. These considerations pave the way for an abductive inference, that also in the real world causal factors underlying the phenomena of interest include those specified in the model. Thus, to validate a model, I must provide a commentary that sets it within a field of inquiry in which certain issues are considered problematic (thus providing the purpose P and the audience E) and set out which parts of the model (m_1, m_2, ..., $m_n \in M$) are the ones that I take to bear resemblance to the target system. If the point of interest is the abductive step (inferring causes from the effects), these must include not only the initial model specification but also the emergent outcomes in the model world. In many cases it is interesting in itself if we can find a limited number of constituent parts that, when set out as a simulation, result in complex, aggregate behaviors; but if these can be set in the context of a theory and fit with empirical findings, we approach a synthesis between the two.

1.2.3 Evolution of Religion on Multiple Levels

While the study of complex phenomena has become established as an interdisciplinary field of inquiry (e.g., Epstein and Axtell 1996; Sun 2006; Miller and Page 2007), the issues involved have been little discussed in the study of religion, with some notable exceptions (Malley 1995, 1997; Bainbridge 2006; Czachesz 2007, 2012). Of these, Malley (1995) argues that recurring social patterns across religious systems speak of self-organization (an emergent property of complex systems) and that they evolve towards criticality, as evidenced by remarkable stability punctuated by dramatic revivals (similiarly Czachesz 2012, 16–18). However, the complex systems view of religion is not so far removed from established approaches as one might assume. Rather, similiar ideas were already expressed by classics of the field, such as Émile Durkheim and Roy A. Rappaport, who both combined (implicit or explicit) system theorizing with an argument for the cohesion-inducing effects of religion. Durkheim (1915) famously suggested that when worshiping gods, members of a society unwittingly worship the idea of their society, creating the necessary social forces for holding it together:

> That is why we can rest assured in advance that the practices of the cult, whatever they may be, are something more than movements without importance and gestures without efficacy. By the mere fact that their apparent function is to strengthen the bonds attaching the believer to his god, they at the same time really strengthen the bonds attaching the individual to the society of which he is a member, since the god is only a figurative expression of the society (Durkheim 1915, 226).

Durkheim's argument is a sociological one, but similiar views have been expressed also by anthropologists (e.g., Douglas 1986, 23–24), who have treated religious ritual as a form of communication. The notion that rituals are specifically structured for establishing social contracts was advanced by Rappaport (1979a, 1999):

> There is surely no intrinsic or causal relationship between dancing and fighting, particularly between fighting in the *future* and dancing in the *present*. However, the *contemporary* state that the transmitter is signaling in this instance is a *conventional* state or act, namely one of promising or undertaking obligation (Rappaport 1979a, 181, emphases his).

In dissecting the "obvious aspects" of ritual that make them different from other forms of symbolic communication, he noted that by virtue of being formal (stylized, repetitive, spatio-temporally recurring), rituals are easily recognizable situations. While there is no guarantee that an individual participating in a ritual has internalized the beliefs it implies, by performing he or she publicly acknowledges the validity of the injunctions it prescribes. Symbolic language allows for considerable subtlety that can leave observers in a state of uncertainty as to what was actually promised, but the relative crudeness of ritualistic displays leaves less room for speculation; ritual dance obligates a dancer to a pre-defined form of help should a war ensue. In addition to his early application of system theory to the Maring ritual cycle (1984), Rappaport (1979b, 1999) himself argued for an evolutionary interpretation of his informatiotheoretical arguments, e.g., that the rise of symbolic language created a new adaptive challenge in the form of expanded possibilities for deception, and that ritual encoding of promises would have diminished this problem sufficiently for human sociability to develop. This can be compared with the almost identical argument regarding the need for "credibility-enhancing displays" to offset the possibilites for manipulation by mental representations in Henrich (2009). Also, Sosis (2003b) has explicitly advanced Rappaport's ideas to account for his findings concerning ritually-induced cohesion.

Similiarly, Durkheim's (1915, 18) views of the social realm as a more complex form of natural reality, and of social contracts as springing from a gradual process rather than rationalist decision-making, accord well with more modern theories of social networks (e.g. Kearns et al. 2009) and emergent properties in adaptive systems (see Pyysiäinen 2005a, 2005b). But his explanation of religion rests on dubious notions of collective consciousness, symbolism, and (super)-naturalness (Pyysiäinen 2003, chapter 4).[8] It seems that concentrating on the social functions religions may have is not possible without taking into account the "supernatural" or "transcendental" content they are often defined as having (*contra* Durkheim 1915, 24–35), as empirical studies (Norenzayan and Shariff 2008) have demonstrated that even in structurally comparable situations, institutions with such content outcompete others when it comes to measures of group cohesion. As Pyysiäinen (2009, 9–12) notes, it is a valid question why precisely those kinds of beliefs serve this function, and he calls for combining "the Tylorian[9] and Durkheimian legacies in explaining how beliefs about supernatural agents are used in organizing a society."

Notably, the Durkheimian functionalist thesis has already been taken up by D. S. Wilson (2002), who proposes that the system of beliefs and moral norms found in religions work to downplay intra-group competition, enabling religious groups to more effectively outcompete other kinds of groups in an evolutionary process of cultural group selection.[10] Wilson's theorizing has received a mixed

---

8 Pyysiäinen (2003, 63–70) argues that contemporary consciousness studies show no evidence of such collective phenomena, that holding god as a symbol of society is problematic since it conflicts with the adherents' beliefs, and that even for "primitive" human beings some ideas are counterintuitive and different from intuitive ideas of natural phenomena.

9 Pyysiäinen (2009, 9) refers here to Sir Edward Burnett Tylor, who was the first to lecture in anthropology at Oxford, and who "found religion in adoration of idols, sacrifice, or belief in a supreme deity". He identifies the neo-Tylorian emphasis on supernatural agents as being often at odds with the Durkheimian/Geertzian cultural systems approach in the contemporary study of religion.

10 Although there has been some confusion on this matter, this proposal is essentially similiar to that of Richerson and Boyd (2005), even if less formally put (Sober and Wilson 1998, 152–156, ch. 5; see Wilson's included commentary to Laland, Odling-Smee, and Feldman 2000). This is in part a continuation of the argument put forth in Sober and Wilson (1998) with the aim of

response, partly as his 2002 argument is aimed primarily against the memetic and rational choice accounts of religion, not taking into consideration the accumulated body of findings of another evolutionary approach, the cognitive science of religion (see Sosis 2003a), although this issue is partly remedied in Wilson (2005) by including a larger empirical test base. While his discussion of how religions in his samples enhanced group coordination is at times anecdotal and does not go a very long way towards settling whether this effect is "incidental" or "fundamental" to them and not to other social institutions, his research program has the commendable quality of taking into account the belief system, behavioral qualities, and socioeconomic context to provide a complete picture of why a given religion has evolved into its present state.

Recently there have been even more ambitious calls for a unified, cross-disciplinary research program that would incorporate economics, human biology, anthropology, sociology, behavioral psychology, cognitive science, and political science via game theory and behavioral experiments into a consistent and synergic evolutionary approach to human behavior and culture (e.g., Gintis 2004; Boyer 2006). Regarding the present state, I agree with Richerson and Newson (2009, 102) that "[t]oday Darwinian social science is perhaps as mainstream as any other variety in this unfortunately fragmented field of inquiry." But it also appears that even the evolutionary approach has become so wide and varied that one may wonder whether there remain any unifying ideas to which all practitioners would subscribe. The school of evolutionary psychology works from the central assumption that the behavior of the modern human can be explained in terms of modular cognitive adaptation to ancestral enviroments and may be maladaptive in modern contexts (Tooby and Cosmides 1992, 114–122). The focus is on the evolutionary time of genetic change, a slow process. This is in contrast to human behavioral ecology (Winterhalder and Smith 2000), where the focus is on human flexibility, which allows adaptation to different environments in the behavioral time of individual decision-making. An approach that combines features from both those aforementioned is the dual inheritance theory of gene-culture coevolution (Boyd and Richerson 1985; Richerson and Boyd 2005), which builds on the insight that the advent of human cultural capacities permanently altered the environment of adaptive challenges and selected for cultural learning and prosocial behavior (with cultural variants being selected by our capacities as cultural learners). Alongside the work of Sober and Wilson (1998) and Henrich (2004), it has resuscitated the idea of group selection (deter-

---

resurrecting "group selection" as an applicable concept in evolutionary biology (see chapters 2.2.2 and 3.2 of the present work).

mined by cultural, not genetic variants), which appears highly dubious to some commentators (Trivers 1998; Houser, McCabe, and Smith 2004).

More specifically concerning the study of religious phenomena, the approach known as cognitive science of religion (e.g., Lawson and McCauley 1990; Whitehouse 2000; Boyer 2001; Pyysiäinen 2003; Barrett 2004) builds centrally on the blueprint of evolutionary psychology as defined above (Visala 2009, 13–14). It has been argued that what we today label "religion" results from our more general cognitive capacities, such as agent-detection and misfiring – seeing "faces in the clouds" (Guthrie 1993; Boyer 2001; Barrett 2004). Long-lasting religious institutions are created by selective retention of cultural representations that are cognitively salient; they activate the innate, evolved modules in our brains and are more likely to be retained and recommunicated than those that are less intuitive and attention-grabbing (Sperber 1996). Most successful are those ideas that are minimally counterintuitive (Atran and Norenzayan 2004): for the most part, these conform to our intuitive expectations (are easy to process and memorize) but make some salient breaches (are striking and capture attention). This leads to the natural conclusion that there is no unified phenomenon that needs to be called "religion," and belief in gods, rituals, and myths can be studied in a piecemeal fashion (Barrett 2007). Religion appears as a by-product of human evolution with little adaptive significance.

However, these conclusions have been contested by other authors who focus on the way religious behavior seems to support group cohesion and prosociality (Irons 2001; Wilson 2002; Bulbulia 2004; Johnson and Krüger 2004; Alcorta and Sosis 2005; Sanderson 2008). Here a central concept is that of a costly signal (as understood in behavioral ecology; Zahavi 1975; Cronk 1994; Bliege Bird, Smith, and Bird 2001; Bliege Bird and Smith 2005), as empirical studies have revealed a connection between demanding religious practices (behavioral requirements and sacrifices) and group longevity or mutual trust (Sosis and Bressler 2003; Sosis and Ruffle 2003; Soler 2008; Bulbulia and Mahoney 2008). Belief in omniscient, moralizing deities appears to have similiar effects (Johnson 2005; Shariff and Norenzayan 2007). As it is uncontroversial that coalitionary behavior (mutuality, trust, group conformity) was at least as important for early humans as it is for moderns (Boehm 2000; Fessler and Haley 2003), or even that "cooperation is always a good thing" from the evolutionary-adaptive standpoint (Tautz 2009, 245), they argue religion to be an evolutionary adaptation. This has lead to an ongoing debate between the two schools of thought (Boyer and Bergstrom 2008; Sosis 2009). It may well be that neither theory has the whole answer, since as Atran and Norenzayan (2004, 714) state, commitment theories lack explanations for peculiar nature of religious commitment and "cannot distinguish Marxism from monotheism," while cognitive theories do not account for the emotional and material attachment to particular deities, and "are unable to distinguish Mickey Mouse from Moses" (see also Gervais and Henrich 2010).

But recent work by Henrich et al. (much of which builds centrally on Boyd and Richerson 1985, and Richerson and Boyd 2005) suggests the gap can be bridged by subsuming the existing approaches within a broader framework of cultural learning and multi-level selection (Atran and Henrich 2010). Minimally counterintuitive ideas have a mnemonic advantage over other kinds of ideas, but people commit to the ideas that are endorsed by those around them, especially if they validate their own commitment to them by performing costly acts (Henrich 2009). Both content and context matter when we observe and learn cultural variants (Henrich and McElreath 2003). Shared commitment to any idea can create a stable state – the group can become "locked" into observing a social norm regardless of its beneficience – and it is the existence of multiple stable states that enables group selection (Henrich 2004). With the advent of agriculture, the selection pressures created by the growth of societies into larger and more complex ones favored groups with religions that had evolved over historical time to having omniscient moralizing deities, elaborate rituals, and prosocial intragroup norms (Stark 1997; Wilson 2002; Roes and Raymond 2003; Marcus and Flannery 2004; Johnson 2005; Shariff, Norenzayan, and Henrich 2009; Henrich et al. 2010). This synthesis – the theoretical context of the present study – incorporates human psychology, cultural institutions, and historical development into a set of interlocking evolutionary processes (operating on genes, ideas, and groups) that have created the patterned variety of religious phenomena observable over time and space.

1.2.4 Norm Groups Emerge from Psychological Rudiments

The synthetic research program outlined above puts major emphasis on the way social context and dynamics of group behavior alter the strategic environment of individual cognitive agents. Adaptive behavior in individuals is dependent on (mostly) reliable expectations of the behavior of the surrounding environment. We humans appear to have fairly dependable intuitive expectations regarding how physical and biological phenomena function (Hirschfeld and Gelman 1994; Medin and Atran 1999). But predicting the behavior of other humans is more complicated; while the expectations and intentions we have do not affect the trajectory of a falling tree or whether a cow produces another cow, an individual supplied with a "theory of mind" (Frith and Frith 2005) may take them into account in his/her course of action. Studies show most people to be "conditional cooperators" who cooperate when they expect others to cooperate and defect when they expect others to defect (Dawes 1980; Gächter 2006; Yamagishi, Jin, and Kiyonari 1999). But these are not straightforward decisions, as the participants have to consider whether the situation demands some particular kind of behavior, whether others involved have come to the same conclusion, and if third parties observing the situation expect certain behavior. What emerges from this is the concept of a social norm, a behavioral rule conditional on expectations we have about the actions of others (see chapter 2.3 and Bicchieri 2006). This conception of norms is similiar to that of an institution within the new institutional economics school (North 1985, 559, 571; 1990, chapter 5); they work to constrain the uncertainty inherent in the otherwise limitless possible forms human interaction can take. Granted, there are forms of cooperative behavior that do not require socially determined norms to support them, but these do not account for the variety of cooperation observable in humans (see chapter 2.2 and Henrich and Henrich 2007, 64–74, chapter 7). Here we see how the conceptual tools at the psychological level suffice to explain how individuals interact with the non-social world around them (the lower levels) but appear lacking when required to account for phenomena emerging from interactions between entities on their own level.

Novel norm environments can significantly alter the behavior of individuals and over time have a substantial effect on their community. It is a property of cooperative behavior that it brings a surplus to those involved; even though individuals incur costs by engaging in it, the total benefit it brings exceeds these. This is what Sosis and Ruffle (2003) found on Israeli *kibbutzim* (see chapter 3.1.4), where daily life is filled with trust dilemmas such as consumption of communal food, water, electricity, and the use of communal cars. The small difference in interpersonal trust in favor of the religious kibbutzniks can accumulate into large effects over time, and this would explain the previous finding of Fishman and Goldschmidt (1990), who noted how the few religious *kibbutzim* show outstanding economic performance in comparison to the secular majority that are staggering under debt. Fishman and Goldschmidt slyly observe how socialist rationalization seems to work only when reinforced by Judaic religion, as "the gap between its [religious kibbutzim's] economic performance and that of the secular *kibbutzim* is so wide as to suggest two different qualitative levels of performance."

Such divergence in fates of communities that differ only in a few crucial aspects recalls the definition presented above: that for a system to qualify as a complex one there has to be interdependence between its parts, and removing one of these parts seriously impairs it, up to the point of destroying the system or its functionality. The studies of ritual signaling give support to the idea that a religious system may indeed be complex.[11] Studying the relation of costly (limiting or sacrificial) behavioral requirements to commune longevity in nineteenth century America, Sosis and Bressler (2003) found that a positive relationship exists, but only when enforced by a religious ideology. The social cohesion required to uphold communal life appears to be dependent on both factors (costly norms and religious motivation), in keeping with the definition of complexity. One quality of such emergent properties is that their causes may be barely discernible on too low a level of inspection, as they result from the accumulation of interactions. For example, the heightened level of trust and cooperativity between co-religionists is almost impossible to discern in a dyadic interaction, except by employing the precision techniques of a psychological laboratory, which reveal a slight quantitative difference when compared to the general population. But in large groups and over historical time this may accumulate to the "different qualitative levels of performance" of the Israeli *kibbutzim*. The data of Fishman and Goldschmidt suggested stronger measures of mutual aid between the religious *kibbutzim*, that whole communities take on an appearance of adaptation and struggle for survival because adaptive micro-decisions made by their members constitute self-similiar behavior, a further property of complex adaptive systems.

1.2.5 Difference to Economist Approaches

Before moving on to the main chapters, one important distinction must be made. The present argument employs several concepts associated with economic thought (e.g., cost and benefit), and in the end claims to resolve issues relevant to human beings' economic behavior. But this relates to a wider entanglement between economics and history of religions. In the preface to his volume on the origins of Islam, Donner (2010, xi–xii) quotes the nineteenth century scholar Ernest Renan to summarize the outlook of antecedent studies:

> We arrive, then, from all parts at this singular result: that the Mussulman movement was produced almost without religious faith; that, putting aside a small number of faithful disciples, Mahomet really worked with but little conviction in Arabia, and never succeeded in overcoming the opposition represented by the Omeyade party.

11 This is not to say that non-religious social systems are not complex, as they most certainly are. Instead, the argument is that components of religion appear to interact with each other as well as with non-religious components of the system so that novel qualitative features that are not present in the non-religious system may emerge.

While admitting that the above position is rather extreme, he claims that twentieth century scholarship has "in subtler guise" reiterated it by reducing the driving force in Muhammad's movement into anything but a religious conviction – social reform, economic stress, nationalism – making religion "merely a pretext for the real objectives." Such views are by no means restricted to the study of Islam. In his study on the Ostrobothnian *Skört* Revival, Ylikangas (1979, 280–282) came to the conclusion that the revival manifested the frustrations of economically and politically deprived social groups. In his mind, it was a religious revival only for the reason that these individuals could not clearly formulate the ends their protest aimed towards. In studies like this, newfound religious motivations of the revived are only the manifestations of latent deprivations created by societal ills (Heino 1977, 36).

However, a survey of the past 30 years of sociological articles (Smilde and May 2010) reveals that this explanatory trend is reversing; whereas three decades ago the primary independent variable was more commonly a social process, today it is twice as often a religious process. Smilde and May take this as evidence of a new "strong" program in sociology of religion that treats cultural processes as autonomous phenomena that are not reducible to social factors but can act as causes and influences on these (and more often than not, religion was found to have a facilitating or enabling effect on some societal function). I present this finding here as it further contextualizes my point that antecedent scholarship on religion has run into serious problems when attempting to explain phenomena that involve religion without taking into account the specific contributions of ritual practice, belief in deities, and other religious paraphernalia to human relations. As both Heino (1977, 36) and Donner (2010, xii) emphasize, it is misguided to ignore the fact that the adherents to these religions were first and foremost concerned with issues of salvation, piety, and righteous behavior. An explanatory strategy that overlooks these factors in favor of more general social concerns can hardly provide a satisfactory explanation for why these people engaged intensely in these practices instead of others. As D. S. Wilson (2005, 430) argues, the question we should be asking here is instead, "What do they cause people to do?" What behavioral patterns do beliefs, narratives, and rituals motivate?

Of course, a classic in this line of argument is the thesis of Weber (1930, 51–53, 79–81, 110–117, 163). He argues that the development of the protestant ethic generated a new economic ethic, the spirit of capitalism, by elevating an earthly calling above monastic pursuits and making salvation dependent on faith and God's grace only. This gave cause to seek signs of this grace, surest of which was success in one's earthly pursuits (as "God blesseth His trade"), but only when carried out in a relentless fashion, regardless of past glories. The thesis has, however, been subject to fierce criticism, including suggestions that the causation is rather the reverse. Lessnoff (1994, 1, 7, 12, 28) and Nelson (2010, 164) state that much of this results from misunderstandings of what the thesis is actually about. For example, Ylikangas (1990a, 335–341) argues that it was the spirit of individual enterprise on the Finnish west coast that also gave birth to individualistic religious expression in the area, e.g., witchcraft in the sixteenth and seventeenth century, or revivalism in the eighteenth and nineteenth. But Kyntäjä (2000, 209–210) points out that Ylinkangas too falls victim here to an overly simplistic interpretation of the thesis in thinking that it presupposes the non-existence of an individualistic or capitalist ethos prior to the advent of the protestant ethic. Weber (1930, 36) himself notes that "a number of those sections of the old [Holy Roman] Empire which were most highly developed economically and most favored by natural resources and situation, in particular a majority of the wealthy towns, went over to Protestantism in the sixteenth century." Furthermore, when Ylikangas criticizes Weber's model for not explaining what gave the protestant ethic its efficacy, he misses the psychological argument in it, that making an explicit connection between a (new) mode of behavior and the fate of one's soul was a powerful motivator for large numbers of people (Lessnoff 1994, 12–14). This point is highly congruent with that of the present study: while religion does not necessarily give birth to ethics or norms that could not otherwise be, it provides the means for large-scale conformance to them.

My aim is not, however, to evaluate Weber's thesis or its applicability to Finnish conditions; I only aim to demonstrate that the present argument is compatible with it, should it be true.[12] Nor am I, in spelling out my opposition to the above-mentioned explanatory reductionism, advocating a diametrically opposite view that the socioeconomic context is of no relevance when trying to explain patterns of religious behavior. On the contrary, as will become evident, I find it to be of utmost relevance. The point I am trying to make is that in order to provide a satisfactory account of a religious process taking place in a specific place and time, we have to appreciate both the contextual factors and the religious aspects. Why did the adherents respond to the posited stressor by engaging in religious behavior, and how did this help them to achieve their objectives, if indeed it did? In the following chapters, I attempt to demonstrate that religion

12 For those interested, Nelson (2010, 168–170, 197–202) reviews some of the criticism and summarizes the contemporary opinion as being that while Weber overstated the effect of Calvinism *vis-à-vis* other forms of Protestantism, taken together their impact on economic development of the Western world has been considerable. The main culprit for the difference between Protestant and Catholic countries he finds in the stagnating effects of the Counter Reformation. On the Finnish case, see Kyntäjä (2000).

is in a unique position to solve social dilemmas precisely because it is *not* about these things, but rather about sincere commitment to a set of beliefs and practices that concern transcendent agents and realities (cf. Pyysiäinen 2010b, 6). To put this very crudely, an assembly of individuals that can commit themselves to a mindset where only one's responsibility in front of one's deity matters – rather than petty grievances they might have against each other, or earthly judicial repercussions, or death on the battlefield (cf. Ibn Khaldûn 1967, 125–126) – can either proceed, in the manner of the *umayyads*, to set up the fifth-largest contiguous empire in world history, or, in the manner of the *Skörts*, stay up all night singing their hymns regardless of the wishes of county constables. But not all religious groups are as successful, and many dissolve outright (Stark and Bainbridge 1985, 126–134). In the course of the present work, I explore why some groups appear more successful than others in fostering commitment and thriving in their socioecological niche.

Another, more specific demarcation is in order. As the present argument will involve the concepts of "cost" and "benefit" in explaining human religious behavior, some may think it a variant of the rational choice theory of religion. According to Stark and Bainbridge ([1987] 1996, 27, 42), its most notable proponents, a core axiom of human behavior is that "humans seek what they perceive to be rewards and avoid what they perceive to be costs." On top of this they build an exchange theory of religion, where "religious organizations are social enterprises whose primary purpose is to create, maintain, and exchange supernaturally-based general compensators" (the term "compensator" being interchangeable with reward/benefit). It may add to the confusion that the work of the economist Iannaccone (1992, 1994) on the relative success of stricter churches is also often cited in support of evolutionary accounts of religion.

Three critiques of these theories are relevant to the present discussion. First, as Ylikoski (2004, 547) has noted, the rational choice theory of religion allows rewards to be anything people might want and thus loses much of its explanatory power. If preferences are deduced from observable choices, explaining human action by human preferences becomes tautological. He concludes that a rational choice theory that makes no substantial claims about what benefits an individual should desire is beyond falsification. Here an evolutionary approach is on stronger ground, as it assumes that it is the benefits that help maximize the reproductive capacity of an individual (or a group) that have molded the human world into its present state, which is a more constrained thesis.

A second, more thorough critique has been spelled out by D. S. Wilson (2002, 49, 81–82), who wonders why, with the rational choice theory's emphasis on goods producable or non-producable by human action, there is no attention given to "the category of goods that can be produced by human action, but only by coordinated human action, and the role of religion in achieving the required coordination" – especially considering that Iannaccone (1992, 1994) attempts to prove that strict demands placed on members by some groups can work to screen out free riders. Others too have observed that "the theme [of religion resolving cooperative dilemmas] is more fully developed by sociobiologists than by rational choice theorists" (Orbell et al. 1992, 292). While Stark and Bainbridge (1996, 84, 113) claim that their "theory conceives of religion as a rational human activity, not as the bursting-forth of deep, irrational impulses," where rationality is marked by "consistent goal-oriented activity," Wilson (2002, 49, 162–165) notes that neither they nor Iannaccone give an actual function to religious activity except to offer "vague psychic benefits" as compensators to unrealisable desires. Indeed, in Iannaccone's (1994, 1183–1188) analysis, the "commodity" produced collectively by members of the religious group is a "religion" that gives out "religious satisfaction" based on how enthusiastically group members perform in the area of religious ritual (singing, reciting, praying, testifying), while just having free riders around participating with less vigor reduces the "welfare" of everyone. He labels the material benefits some groups produce as "externalities" and does not seem to be very interested in them. This seems odd, as economists have long appreciated the problems inherent in production and usage of collective goods (Olson 1971; Hardin 1968). And indeed, they concede that "when commitment levels are high, groups can undertake all manner of collective actions and these are in no way limited to the psychic realm;" one example of this would be rebuilding co-religionists' houses (Stark and Iannaccone 1997, 145). It seems to me that Iannaccone's argument is aimed at explaining why people choose to join strict sects *rather than* liberal churches, despite the fact that they both supply religious or spiritual experiences, which would (apparently in his mind) be indistinguishable in content were it not for the dilution by free riders. He does not even attempt to answer the question of why there is a demand for "religious commodities" in the first place.

The third problem is that of the "rational choice" in general. Instead of matching costs with benefits, a more cognitively grounded approach to human behavior will have to address biases and dispositions that on average have proved successful. Our behavior is not driven by rational reflection on advantages and risks but by contextual cues and heuristics; for example, we cooperate if we think others will, and even "cheap talk" can contribute to this (Fehr and Fischbacher 2005, 165–167, Sally 1995, 78; Bicchieri 2006, 144–145). A recent cross-cultural survey of 15 small-scale societies concluded that there was no society in which experimental behavior was "even roughly consistent" with the canonical model of the purely self-interested (rational) actor (Henrich et al. 2001; Henrich et al. 2004, 5, 49–50). There are also studies documenting an increase in prosocial behavior when observation is even hinted at. When a photograph of a pair of eyes (no head, face, or body) was placed above an honesty box for payments of tea, coffee, and milk in a naturalistic setting (a workplace where the system had been in place for years), payments increased massively when compared to a control condition, an image of flowers which was used on every second week (Bateson, Nettle, and Gilbert 2006; see also Haley and Fessler 2005 and chapter 3.1.2). It seems that even small, subconsciously registered hints of observing agents being present lead us to behave more cooperatively towards others. Such results point towards a rather different conceptualization of "rationality" than what is assumed by the rational choice theorists.

1.3 Research Questions and Methodology

Ecologists distinguish between the evolutionary timescale of genetic change (a slow process) and the behavioral timescale of individual decision-making (a faster process) in discussing dynamics of ecological systems. In between these two we can situate the historical timescale of cultural change, in keeping with the above perspective of evolution on multiple levels. But sometimes the pace of historical change is so slow as to be imperceptible to a human mind, while at other times it occurs rapidly so that individuals can sense they are living in tumultous times (become aware of "emergent features," cf. Gilbert 1995, 8). In the latter case, this awareness interferes also with behavioral decisions. In systemic terms, the social system evolves towards criticality where a single event, previously insignificant, can cause system-wide change. Such a stochastic event can cause the system to develop an emergent property of which the agents become aware, and they adjust their behavior accordingly. This may or may not further fuel the change (a feedback loop). Concerning religions, we can recognize this kind of dynamic in the rhythm of stable socioreligious systems punctuated by revivals that challenge the religious identities of those within the system (cf. Malley 1995, 18; Czachesz 2012, 16–18). In this view, we should not expect revivals to be initiated by a charismatic personage, even though they may take the lead in an already emerging movement. The ideas embraced by revivalists may not be so dramatically different from what has been expressed before, but they have suddenly become topical for many. The revivals appear symptomatic of long-term environmental changes that have gradually piled up. And we should have considerable difficulties in predicting their occurrence, as no single factor or event has to be of determining importance.

My study addresses this dynamic of interdependent cultural and behavioral change. Based on the ideas presented above, I suggest that religious revivalism can be understood as a systems-level adaptation to changes individuals perceive in their environment. In evolutionary terms, it forces individuals into a new situation where the contextual factors that influence the learning of cultural variants have been altered, as the commitment of those around to them is fluctuating. It also allows for the reappraisal of their content, as long-held variants *vis-à-vis* more novel ones may resonate in varying degrees with the other ideas existing in the now-current cultural environment. Widespread adoption of novel variants may initiate a group-level selection process, as the revivalists "lock" onto their new variants, creating their own norm environment and bringing into being a new social system. The resulting competition between cultural groups may or may not be violent, as the more adaptive group may also subsume the other by attracting converts or by exposing the superiority of its norms, resulting in a quiet takeover (Henrich 2004, 28–29). These preliminary assumptions lead me to consider the following questions:

1. *Who were the revived?* The position of an individual in the system is determined by multiple factors relating to societal status and life history. These factors also influence the operation of the heuristics an individual applies in determining his/her course of action (e.g., the contextual bias of imitating those who appear most similiar to oneself; Henrich and McElreath 2003, 129). The events leading to the revival may also affect those involved differentially, if some parts of the population are more dependent than others on an environmental resource.
2. *What were the experiential factors?* The individual parts of the system do not observe the whole system. They may become aware of some emergent features in it, but even then they rarely accurately discern its causes. The factors that trigger individual behavior must be ones that are available to the individual experience – the possibilities and constraints of daily life. This is not to say that the individual must be aware of these factors, as studies show that even subliminal cues can affect behavior in ways conscious reflection would not approve of; for example, images of eyes appear to prime a feeling of being watched, leading to more prosocial behavior (Haley and Fessler 2005; Bateson, Nettle, and Gilbert 2006; Burnham and Hare 2007).
3. *How did the system self-organize in response?* An adaptive system can respond to changes by introducing novel arrangements in its micro-level interactions. In order to understand the process the system is undergoing, we must seek to identify how the emerging self-organization addresses the adaptive challenge. In the case of revivals, this often takes the form of introducing new social norms that demarcate the in-group from the out-group. In Luhmann's (2006) systemic terms, what is happening here is a partial discontinuation of communication, as revivalists may be unwilling to continue

> some of the forms of communication that are prevalent within the wider societal system, while some of their communicative signals (such as outlandish dress or stringent behavioral requirements) may be incomprehensible to outsiders.

In answering these questions, my method is principally to seek correspondence between the theoretical and the empirical by systematic exploration and analytical argument. Concerning the theory on one hand, I argue that existing empirical studies and theoretical thought allow us to build a synthetic picture of religious groups as cohesive norm units and extend this into accounting for religious revivals. I support this argument by demonstrating its plausibility with a simulation experiment. As for the empirical side on the other hand, I seek to expose how the revivals of Christian belief in early nineteenth century Finland exhibited many of the features predicted by the theoretical construct elaborated. To this end, I will review sources and studies concerning three revivals: the southwestern Jumping Revival (*Satakunnan hyppyherätys*), the Ostrobothnian *Skörts* (*Pohjanmaan körttiläisyys*) and the Kuortane Revival instigated by Jakob Wallenberg and later led by Isak Vasu.[13] This selection is not arbitrary but is dictated by extant source materials: these are the only revivals for which we have systematic data on their social composition due to mass court processes or inspections for violating the conventicle placard. There were also other processes, but these involved only those accused of organizing the conventicles, not all participants, and after 1844 the placard became a dead letter, so the Laestadians eluded its grasp completely (Pentikäinen 1978, 197–200).

While one might (on various grounds) argue that these three revivals are not representative of the diverse revivalist activity of the period, I reason that this se-

---

13 On choice of nomenclature: There was also another, contemporary "jumping revival" at Karelia, but these two are unconnected (Puolanne 1936; Heino 1976, 66–70). I use the name exclusively in reference to the southwestern revival. Regarding *Skörts*, a customary English designation for them appears to be "pietists," but this is misleading in the early nineteenth century context since the historical association between the early revivals and the German Pietist movement is dubious (see ch. 5.1). Thus I have chosen to use the appellation they received based on their clothing (see ch. 5.3.3) and have excluded the Savonian branch according to the dictates of available source material.

Regarding the last-mentioned point although both nineteenth century sources and later scholarly works designate this movement as "followers of Vasu" or "Vasuism" (*vasulaisuus*), or speak of "Wallenberg's teaching," I have preferred the term "Kuortane Revival" as my focus is on this parish in 1799, eleven years before Vasu took the lead, and 1825, at which point it would seem excessive to attribute everything to Wallenberg's 1798 goings-on at the Kauhava parish. For more details, see chapter 5.3.2.

lection criterion does not presuppose my theoretical hypotheses in any way (cf. Wilson 2005, 424). Rather, at the outset I lamented that this criteria precluded an analysis of the highly emotional practices of the Laestadian Revival, while forcing me to consider the supposedly more sober Ostrobothnian Revival, with the additional complicating factor of clerical leadership. And while one might consider the possibility that such mass processes were deemed necessary only in areas where the revivalist activity reached exceptional heights, such a view is not supported by the research; in many other places the revivals were also seen as deeply disturbing to the established way of life (e.g., Raittila 1976, 24–25; Hartikainen 2005, 206–214), which often led to more limited persecution. More likely it was simply the question of local authorities' enthusiasm and the high church authorities' circumspection that in some places lead to more wide-ranging processes than in others (Pentikäinen 1978, 193–195; Ylikangas 1979, 17–18). That said, I will occasionally draw also on conclusions made by scholars regarding other revivals, especially when this is necessary to pre-emptively counter some criticisms that might arise. The systematic analysis, however, is restricted to the three aforementioned revivals.

For my espousal I have reviewed original documents contained in the seven-part collection *Historiska Upplysningar om Religiösa rörelserna i Finland i äldre och senare tider* by Matthias Akiander, which constitutes perhaps the most important publication concerning nineteenth century religious movements in Finland (Huhta 2007, 24–25). Material pertaining to the revivals under scrutiny is to be found in parts III (1859), IV (1860), and VI (1862).[14] Murtorinne (1986, 132), while finding some of Akiander's own analyses lacking, lauds his "remarkably wide-ranging and objectively compiled documentary material." Ruuth (1942, 136–137) also considers Akiander's work reliable, but notes that it has some deficiencies, most notably concerning Ostrobothnian *Skörts*. Apparently this resulted from mistrust among them regarding his historiographical approach and also from a desire to cover up still quite recent rifts within the movement (Rosendal 1915, 140–176).[15] This lack is somewhat compensated by Mauno Rosendal's four-

14 Later a professor of Russian language and literature, Akiander began this work as a scholarly history of religious movements in Finland, for which he copied and collected numerous documents from archives, personal letters, and memorates from "clergy and respectable citizens" (Ruuth 1942, 133–135). But as the amount of material grew, he made the decision to instead publish much of it in its original form, as a collection of source material for other scholars to use.

15 J. J. Rahm, a Savonian *Skört* minister, complained in a May 2, 1862 letter that "Paavo Ruotsalainen was God's chosen instrument, who had the profound secret of the Gospel hidden under a coarse shell, and Akiander is an honest, insouciant [not of the revived] man, who understands the secret of godliness as well as a cow does a windmill's. Is it any wonder then, that A. defames

part history *Suomen herännäisyyden historia XIX:llä vuosisadalla* (1902, 1905, 1912, 1915), which, although written from an insider's perspective, contains valuable material not found elsewhere (Huhta 2007, 26, 28). All English translations of source material included in the present work are by myself, made from the Finnish or Swedish original. Concerning translation of vocational terminology, for the most part I have used those suggested by Greus and Hirvelä (2006) or Ylikangas (2002). However, I have chosen to translate *talonpoika* as "householder" instead of "peasant" or "farmer," since for the purposes of the present argument it is important to emphasize the distinction between those of the land-tilling group who owned their houses and fields, in contradistinction to different kinds of tenant farmers.

My analysis also makes much use of church and sociohistorical literature. The standard reference on the Jumping Revival is still Heino (1976); even the critical reappraisal in Sulkunen (1999) relies heavily on his work. The Kuortane Revival is little studied, but Troberg's 1995 Masters thesis contains both a review of what literature exists and an analysis of previously unpublished archival material. In contrast, volumes have been written on the Ostrobothnian *Skört* Revivals, but many of these works emphasize the role of leaders (Huhta 2007, 27; e.g., Remes 1995) or relatively minor doctrinal differences (Hartikainen 2005, 275; cf. Ruuth 1942, 142). As these are not central themes in the present work (but see chapter 7.2.1), I have consulted mostly works by Ylikangas (1979, 1990b) and Hirvonen (2003), which take an explicitly sociohistorical perspective. Unfortunately, very little has been written on this subject in English, the only publication highly relevant for the present work being by Ylikangas (2002).

---

and speaks ill of Paavo Ruotsalainen? He must do so; one must not wonder about it, 'it is his task'" (Rosendal 1915, 146).

2 Cooperation as a Human Puzzle

> The error of storytellers results from the fact that they admired the vast proportions of the monuments left by nations of the past, but did not understand the different situation in which dynasties may find themselves with respect to social organization and cooperation. They did not understand that superior social organization together with engineering skill made the construction of large monuments possible. Therefore, they ascribed such monuments to a strength and energy derived by the peoples of the past from the large size of their bodies. But this is not so [...] (Ibn Khaldûn 1967, 144)

2.1 Ubiquitous Cooperation

The fourteenth century social theorist Ibn Khaldûn (1967, xi, 106–115, 134–137) observed how established dynasties of his native Maghreb grew weak, as sedentary life made successive generations take more and more of their position for granted and become lax. At this point (the fourth generation, in his mind) they were soon toppled by another, stronger group from the wilderness that had not lost its vigor – beginning the cycle anew. For him, the central explanatory variable was the "group feeling" or "solidarity" (*'asabîyah*) that "produces the ability to defend oneself, to offer opposition, to protect oneself, and to press one's claims," and which accounted for the difference in scale of cooperation among peoples of the world.

But while recognizing the utmost importance of cooperation, Ibn Khaldûn is less specific on what – or if it is luxury and laxity, how – causes the want of cooperation. More recently, game theorists have appreciated it as the fundamental problem of social life, that groups survive best when their members provide for each other, while individuals survive even better when they take what others provide and give nothing in return. Thus, despite the advantage cooperative action would often confer to participants, it may not appear in everyone's immediate interests to achieve it. Rather, in many situations it is most advantageous for a single individual to refrain from cooperating (defect) when others are investing in the common good. In this case he or she freerides on the efforts of others, not incurring any personal costs but participating in the common benefits. An honesty box system for workplace coffee can be maintained if almost everyone who drinks the coffee pays for it, even if a few freeriders drink without paying. By coordinated action, individuals can kill much larger game, but every individual carries the risk of being maimed. Enviromental change may be contained if most or all major polluters participate, but a single polluter can get away with defecting and outcompete others by refraining from investing in cleaner technology. And if all of the soldiers on one side of the battle choose to lag behind the others for personal safety, they all get slaughtered. Thus, these situations present social dilemmas for the participants, as it is not easy to state outright how they should behave given the options. But to facilitate their study, they can be formalized as game theoretical constructs, the most famous of which is the prisoner's dilemma.[1]

Still, despite this difficulty, cooperation and other prosocial behavior[2] abound; friends help each other, people recycle and donate to charity – but not everywhere and every time. How do individuals decide when, how much, and with whom to cooperate? There are (at least) three levels to this question (cf. Henrich and Henrich 2007, 36). The first of these is the psychological level, where we ask how various psychological factors such as cognition, emotions, and personal dispositions come together to produce observed instances of decisions to cooperate; the second is the institutional level of cultural practices such as norms, technologies, or legal enforcement that can alter objective payoffs and individual preferences; and the third is the level of genetic evolution that explains how humans came to have the required capacities, both as direct psychological adaptations and as cultural learning biases. In the present chapter, I

---

1 The name refers to a hypothetical scenario, where two accomplices face prosecution. If they remain silent, they get minimum sentences (six months). However, both are offered individually the chance to confess and go free while implicating the other, who gets a ten-year sentence. But if both confess, they are both nailed for five years. As a game-theoretical construct, the conventional form (e.g. Axelrod 2006, 206–207) involves two players, who both face a choice of whether to cooperate with the other or to defect. They make their choices simultaneously, after which they receive their payoffs. If both cooperated, both receive the reward for mutual cooperation (*R*). If both defected, they receive punishment for mutual defection (*P*). In cases where one player chooses to cooperate while the other one defects, the defector will receive temptation to defect (*T*) while the cooperator receives the sucker's payoff (*P*). To ensure that defection is more tempting than cooperation, but also that the common good exceeds total benefits for defecting, values for the payoffs should satisfy following conditions: $T > R > P > S$ and $R > (T + S)/2$. But this means that defection strictly dominates cooperation, that is, no matter what the other player does, one will gain more by defecting than by cooperating. This makes the prisoner's dilemma a conservative tool for examining the emergence of cooperation, but see Cosmides and Tooby (1989, 54–55) for a rationale as to why many natural situations resemble this dilemma.

2 There are various definitions for these terms. I follow the usage of Henrich and Henrich (2007, 37–39) who define "cooperation" as occuring when "an individual incurs a cost to provide a benefit for another person or people." Costs can be time or resources, and in complex societies benefits can include upholding some public institution such as voting, food-sharing, or recycling. In this context "altruism" is usually defined in the same way, but I employ it less as a technical term and more in keeping with the everyday usage of concern for others' welfare. "Prosociality" refers to all social actions directed towards upholding the welfare of others or a society, including punishment of individuals who attempt to benefit without paying any costs. These latter are termed "defectors" or "free riders".

briefly introduce the current lines of inquiry to research on the psychological and evolutionary levels. The remainder of the work will be mostly concerned with expounding how the institutional level can interact with these two.

2.2 Evolutionary Considerations

From an evolutionary perspective, cooperation is almost always a good thing for those involved (cf. Tautz 2009, 245). Recently it has become commonplace to emphasize that evolution is not only about "nature, red in tooth and claw" (Lord Tennyson) but also about love, nurture and commitment (see e.g., Sober and Wilson 1998; Nesse 2001; Roos 2009). Fessler and Haley (2003, 8) write: "Today cooperation is arguably one of the most important determinants of human survival and success, and this is likely to have been even truer for that vast majority of our species' history when we lived as nomadic hunter-gatherers." To them and many others, "this suggests that natural selection will have favored psychological attributes that enhance the individual's ability to engage in, and profit from, cooperative enterprises."

Boehm (2000, 79, 82, 84–85) names bullying behavior as a likely to have been the first behavior that was thought to be morally deviant and thus requiring sociocultural control, as extant hunter-gatherer bands are politically egalitarian despite the innate tendency in humans and other primates towards forming hierarchies. But "niceness" in this view is relative. As Sober and Wilson (1998, 9) note, "[g]roup selection favors within-group niceness *and* between-group nastiness". Competition for survival is not absent in this approach, and the adaptations that support cooperation for survival may not be very benevolent in their proximate guises (e.g., punishment). Also, every model of organisms cooperating contains the possibility of invading defectors outsmarting the cooperators by partaking when something is offered but defecting on reciprocation. The puzzle is, how then did humans evolve to cooperate on such a massive scale?

2.2.1 Kin, Reciprocity, Reputation

Natural selection can operate if there are statistically reliable patterns in the environment. In the case of cooperation, this means that performing costly acts that benefit others must anticipate getting at least as much in return. This can be expressed as a simple formula: $\beta > c/b$, where c is the cost of cooperation, b the benefit, and β is the regression coefficient (parameter), which can measure different things depending on the nature of cooperation (Henrich 2004, 4–6).

Thus, any solution to the problem of cooperation requires maintaining sufficiently high value of β. Nowak (2006) and Henrich and Henrich (2007, 43–64) show how this applies to several possible mechanisms. For example, according to the theory of *kin selection* or *Hamilton's rule*, natural selection favors cooperation if relatedness between the individuals[3] r, substituted for β, exceeds the cost-benefit ratio. In this situation, the cooperators are to an extent helping their own genes spread. For empirical examples, in a work that combined ethnography with behavioral experiments, Henrich and Henrich (2007, 93, 96–97) found the Chaldean (Iraqi Catholic) population living in Detroit to be well aware of who their close family members were, and noted that they restricted the costliest forms of cooperation (help with bills, illness, or housing) to those with relatedness of 0.23 to 0.43 – despite their own claims of being "one big family."

However, human cooperation does not only occur between relatives. Axelrod ([1984] 2006) claimed that if interaction between two individuals is frequent, *direct reciprocity* explains the evolution of cooperation, as cooperating now may mean continued mutual cooperation, which is better than continued mutual defection. This requires the likelihood for future interactions with the same partner w to exceed the cost-benefit ratio. In a computer tournament of the iterated prisoner's dilemma game (Axelrod 2006), the strategy of simple reciprocity ("tit-for-tat" [*TFT*] or some slightly more forgiving variant) was deemed superior. Since then, it has been shown that there may not be any evolutionarily stable pure strategies in an iterated prisoner's dilemma (Henrich and Henrich 2007, 48–51). This does not mean that direct reciprocity is not observable in human life, but rather that we should observe reciprocity-based cooperation primarily in dyadic relationships where cultural institutions limit the choice of strategy. Henrich and Henrich (2007, 51, 118–119) report this to hold with the Chaldeans, who for example have a funerary tradition of donating money (to cover expenses) in signed envelopes, so the recipient knows who to remember when the time comes.

---

3 Following the probability of sharing the same genes by descent, for diploids like humans r is 1 between identical siblings, 1/2 between parents and their children or between non-identical full siblings, 1/4 between grandparents and grandchildren, 1/8 between cousins, etc. Interbreeding and non-Mendelian inheritance mechanisms can complicate this. The significance of this dilution of relatedness was appreaciated already by Ibn Khaldûn (1967, 99): "It means that pedigrees are useful only in so far as they imply the close contact that is a consequence of blood ties and that eventually leads to mutual help and affection. [...] This means that when common descent is no longer clear and has become a matter of scientific knowledge, it can no longer move the imagination and is denied the affection caused by group feeling. It has become useless." In practice, humans recognize close kin from multiple cues, most importantly that they "hang around mom" (Henrich 2004, 7–8).

In groups cooperation is often asymmetric and does not follow simple reciprocity. This can be explained as *indirect reciprocity* made possible by reputation: cooperating causes third parties to consider us trustworthy, and increases the possibility they will act cooperatively towards us. Information on our past actions is often spread by means of gossip, and β becomes q, the probability of knowing someone's reputation. In the Chaldean network, reputational information spreads through gossip, and reputable clients are given credit, something that does not initially apply to non-Chaldeans for whom there is no easy method with which to check their reputations (Henrich and Henrich 2007, 123–124). While these forms account for many cases of altruism observable in the natural world, the scale of human cooperation cannot be explained solely by them. Although among humans the intensity of cooperation among kin is greater than among non-kin, genetic relations clearly do not account for everything.[4] For reciprocity, real-life social situations introduce several complicating factors. Individuals may occasionally perceive cooperators as defectors and vice versa (Greif 1993, 530), which creates a trade-off between retaliatoriness and generosity of a strategy (Henrich and Henrich 2007, 51–53). More importantly, group size can be a serious inhibitor. Axelrod's (2006) model of dyadic interactions does not generalize for larger groups, as a formal model by Boyd and Richerson (1988) shows that in a population of defectors and conditional direct reciprocators the probability of forming random groups of cooperators decreases geometrically with increasing group size, if no additional incentives are considered (for a similiar earlier argument see Olson 1971, 45–61). And indeed, a survey of empirical studies reports that "all experimenters who have made explicit or implicit comparisons of dilemma games with varying number of players have concluded that subjects cooperate less in larger groups than in smaller ones" (Dawes 1980, 186).

Modelling work gives more support to reputation-based indirect reciprocity maintaining cooperation (Nowak, Page, and Sigmund 2000; Panchanathan and Boyd 2003, 2004). Also an experiment by Engelmann and Fischbacher (2002) is consistent with this prediction; in a helping game individuals were more likely to help those they knew to have been helpful themselves in the past, and this was further modulated by whether they knew their own action to be public or not (with less helping in the latter condition). However, we can still find exceptions, that is, cooperation in situations where reputational gains

4 This also goes against the "big mistake" hypothesis that human cooperation results from mistaking the vast contemporary social worlds for small, interrelated hunter-gatherer societies of the evolutionary past (Cosmides and Tooby 1989, 57; Richerson and Boyd 2005, 229–231).

are absent (Fehr and Gächter 2002). And while it can in some cases be argued that we operate following catchall heuristics that do not apply well to anomalous situations (such as those introduced in behavioral experiments), the highly context-sensitive reputation formation does not seem a likely candidate for that. It appears more likely that the solution to the puzzle of human cooperation is somehow tied to our also rather unique cultural capacities (Henrich 2004, 30–31), as the scale of human cooperation has grown in magnitude during the cultural history of many societies (McNeill and McNeill 2003).

2.2.2 Cultural Group Selection

The puzzle of human cooperativity can be solved by taking into account that natural selection does not act only on individuals but also on groups – a hypothesis termed *group selection*. If group members behave cooperatively, i.e., in a way that is individually damaging but benefits the whole group in excess to these costs, this group outcompetes those not engaging in such behaviors, and selection will favor cooperation. In evolutionary biology, group selection theories have been controversial to the extent of being considered invalid (see Sober and Wilson 1998, chapter 2; Trivers 1998) or at best confusing (West, el Mouden, and Gardner 2011, 246–250). The reason for the former view is that for group selection to be viable, selection pressures between groups would have to be stronger than pressures inside the group (between individuals). As between-group migration dampens genetic group differences, this becomes less likely: β is zero if groups are randomly mixed in every generation, and negative if defectors can preferentially enter groups of cooperators (Henrich and Henrich 2007, 216). However, proponents of group selection on cultural traits claim it does not suffer the problems of group selection acting on genes (Henrich 2004, 16–17, 22–26; Richerson, Boyd, and Henrich 2003, 364, 368–369; Sober and Wilson 1998, 149–154; Cronk 1994). The main reason for this is that dynamics of sociocultural interactions can create multiple stable equilibria in a group situation and thus maintain between-groups variation, enforcing a high value of β (cooperators interacting with other cooperators) in some groups. In less technical terms, this refers to the variety of social norms that a human group can adopt as its own, cause newcomers to conform to, and the differential consequences of these norms with regard to the group's cohesion and success. This introduces stable between-group differences on which group selection can act.

As stated above, there is confusion over whether the group selection theories constitute competing explanations to the individual-centered approaches. West, el Mouden, and Gardner (2011, 234, 249) state that it is only a question of doing the mathematics differently: whether costs and benefits are calculated within the more conventional inclusive fitness theory or a multi-level model that separately tracks the contributions of group and invididual levels. Henrich (2004, 17–19) offers an illuminating example of this by demonstrating how the superiority of "tit-for-tat" strategy over pure defection can be analyzed also in group selectionist terms, and that this superiority is driven by the between-groups component, not the within-groups component.[5] He notes that the viability of reciprocity is not usually calculated in this way because it is simpler to derive relative fitnesses by averaging the results across all groups (opposing strategies), which West, el Mouden, and Gardner (2011, 240) claim can be done with all evolutionary explanations depending on group selectionist thinking. Sober and Wilson (1998, 32–33) argue against this "averaging fallacy" because in their mind it defines group selection out of existence and conceals the separate group and individual level processes, while West, el Mouden, and Gardner (2011, 249) criticize group selection jargon for obscuring the underlying mechanisms, such as ostracism and reciprocity. In the present work, while I rely to an extent on the cultural group selection framework of Richerson, Boyd, and Henrich (2003) and its theoretical predictions, I am (rather than obscuring) centrally focusing on the proximate mechanisms that regulate cooperative behavior in religious groups.

In the example of Chaldean society (Henrich and Henrich 2007, 146–147, 196), fear of bad reputation enforces a number of social norms, such as endogamous marriage and care for elderly parents, behaviors only predicted by the cultural group selection approach. And predictably, they also preferentially interacted in social and professional relationships with those who were seen as Chaldeans, thus satisfying the condition for having a "high β". Henrich (2004, 22–26) enumerates four propensities of human psychology that are responsible for these behaviors (and provides evidence for them):

- *Conformist transmission.* Humans have a bias for copying behaviors that occur at high frequency among those around them. This is because in many circumstances individual learning would be too time consuming, so it is more adaptive to be a social learner and adopt those behaviors that many others already practice, since they are likely to be workable.
- *Prestige-biased transmission.* A propensity similiar to conformist transmission is to imitate those individuals who are most successful, as this is a good index of practicing behaviors that are functional. The most salient

5 The reason for this is that the within-groups component is always negative because *TFT* always loses the first round to pure *D*. Reciprocity is only favored when the number of rounds played with the same partner is high enough for paired *TFTs* to overcome this loss – that is, they must be embedded in stable groups for this to happen.

cue for success is the prestige an individual is granted by others, but as it is often difficult to say just what he or she is doing right to achieve such a status, social learners also end up copying behaviors that have nothing to do with succeeding. However, often this needs to be tempered by self-similiarity bias, as it may be misguided to copy individuals whose abilities and cirumstances differ too much from those of the learner.
- *Punishment of non-conformists.* While punishment in itself is insufficient in upholding prosocial behavior (see below), mathematical modelling (Henrich and Boyd 2001) shows that it is possible for the punishment of those who fail to punish defectors to stabilize at some level of recursive punishment. As defectors become increasingly rare with the order of defection, the costs of punishing become negligible, and preference for conforming behavior overcomes preference for higher payoffs. This is dependent on a pre-existing conformist transmission bias, but once it exists even in a weak form, these two propensities can plausibly ratchet up each other.
- *Normative conformity.* Individuals in this case try to match their behaviors with those already common in their group. This differs from conformist transmission in that the aim here is not to adopt the most beneficial strategies but to avoid standing out from the crowd and being punished. This may result in their outward behavior differing from their internally held beliefs, but it is the behavior that others observe and learn from.

The concept of group selection implies competition between groups, and violent conflict that disbands one group is one form it can take. But there are at least two others: demographic swamping and prestige-biased group selection. The former refers to one group outbreeding the other. In the latter case the superior norms of one group cause the members of another to copy their behaviors, assuming the intragroup norms of the copying group do not prevent this (even extra altruism can be the target of punishment; Herrmann, Thöni, and Gächter 2008). This allows also for recombination of cultural forms to create even more efficient variants. Based on empirical cases, Henrich (2004, 28–29) suggests that these three mechanisms operate on different timescales: demographic swamping over millenia (as in agriculturalists replacing hunter-gatherers), direct competition or conflict over 100–1000 years (see Soltis, Boyd, and Richerson 1995, 481–482), and prestige-biased selection even over decades. Cases of group-beneficent cultural trait proliferation have also been documented by Soltis, Boyd, and Richerson (1995), Stark (1997) and Wilson (2002).

Cultural group selection enables also genetic evolution, as the spread of novel cultural variants changes the selective environment, and those best equipped for the new cultural situation will thrive. The least controversial example of this happening is the evolution of lactose tolerance in some human populations,[6] but Richerson and Boyd (2005, 196) and Henrich (2004, 30) argue that it is especially human social capacity that is best viewed as resulting from such a process, as in a social environment gaining social rewards and avoiding punishment becomes a priority if one is to avoid becoming an outcast. The centrality of punishment and the avoidance of punishment for our prosocial tendencies is currently a debated issue (Sigmund 2007; DeScioli and Kurzban 2009). The problem is that punishing someone is in itself costly and thus a cooperative act; it benefits mostly those members of the group who are not being punished and do not participate in the punishing. This creates a second-order social dilemma, as everybody should wait for somebody else to do the punishing. However, an experiment (Fehr and Gächter 2002) indicates that allowing for costly punishment actually creates stable and increasing levels of cooperation. Also cross-cultural evidence (Henrich et al. 2006) demonstrates that costly punishment is reality.

Modelling work by Boyd and Richerson (1992, 182–183) and Henrich and Boyd (2001) show that it is possible for the punishment of those who fail to punish defectors to stabilize at some level of recursive punishment, and that punishment can stabilize *any* behavioral norm regardless of whether there are any benefits to it (which is again a situation in which cultural group selection can favor those groups that embrace more beneficient norms).[7] But empirical evidence for punishment of *n*th order defectors remains to be had. Some conceptual clarifica-

---

6 Lactose tolerance is controlled by a single gene, and a history of keeping cows and consuming fresh milk is the best predictor for a high frequency of this gene in a population (Richerson and Boyd 2005, 191–192). Thus, dairying (a cultural practice) created a selective pressure that favored lactose tolerant individuals (genetic solution). But among those peoples that also developed the practice of turning milk into mostly lactose-free cheese and yogurt this pressure was removed, and the tolerant gene did not spread. Here, the different paths adaptive cultural evolution has taken have clearly also influenced genetic evolution (Henrich and Henrich 2007, 31).

7 It is important to note that strong norm conformance does not automatically mean adaptation. The cross-national study by Herrmann, Thöni, and Gächter (2008) revealed that in some present-day countries (mostly from Eastern Europe and the Middle East) people punish those who behave more cooperatively than they themselves have, with unfortunate consequences to the net results. Both this antisocial punishing and the more prosocial form were correlated with the norms of civic cooperation in these countries. The result testifies to the variability of norms that sanctioning can uphold. And as environments change, even previously adaptive norms can become maladaptive. For an illustrative example, Oliver and Atmore (2001, 24–25) describe how the steppe etiquette of the so far succesful Turkish cavalrymen on both sides of the fifteenth century Ottoman-Mamluk conflict led them to reject the use of firearms. The balance was tipped by the Ottoman janissaries (conscripted from Balkans) who had no such scruples and became a highly efficient institution unto themselves.

tion is also needed to differentiate between direct costly punishment and indirect punishment through reciprocal defection, gossip, and ostracism (cf. Sigmund 2007, 598–599). In the latter form, punishment is present in standard models of direct and indirect reciprocity, and Ohtsuki et al. (2009) demonstrate that in this case there is only a small parameter region in which additional costly punishment is an efficient strategy. Notably, Henrich and Henrich's (2007, 123–124) Chaldeans were more concerned with avoiding those with bad reputations than seeking out those with particularly good ones. It may thus be that when direct peer punishment is employed in real life, it serves some other function, such as that of advertizing personal qualities of cooperativeness (Fessler and Haley 2003, 22) or retaliatoriness (Frank 1993, 162–164).

2.2.3 Costly Signaling

One more mechanism of cooperativity needs to be introduced, because an ambitious theory of ritual signalling has been built upon it (see chapter 3.1.4). As has been iterated above, an individual often has the incentive to give a false impression (a signal) of his or her intentions or abilities in order to trick another into committing. The receiver of such signals faces the problem of determining their truth value. The theory of costly signals, or the "handicap principle," known in evolutionary biology is aimed at explaining how traits such as a male peacock's tail, which initially seem to be all costs and no benefits, could function as a signal of fitness that females can rely upon because its costliness makes it less likely to be faked.[8] But the idea that signals can be honest because they are costly to produce has an even longer history in economics, as Veblen ([1899] 2004, chapter 4) discussed conspicuous consumption of luxury as a signal for reputability. Since then, Spence (1973) has argued that level of education works as a reliable signal of employee ability (for those with less ability, university education is comparably more difficult to achieve). He underlined the critical assumption that:

---

8 Zahavi (1975) argued that while female preference for such a trait creates a new selective pressure favoring it, in order for the preference to become more common it also has to offer some relative advantage. This happens if the cost of growing the tail is negatively correlated with fitness, i.e., the males with the more exaggerated trait are also more fit. The tail becomes a reliable signal of fitness that females can trust. This does create a tradeoff where advertizing one's quality actually lowers it and forces females to compromise in their choices. However, if the fitness of their offspring is improved more by this informed choice than it is handicapped by having the exaggerated trait, it – and the preference for it – should become more common.

> [...] a signal will not effectively distinguish one applicant from another, unless the costs of signaling are negatively correlated with productive capability. For if this condition fails to hold, given the offered wage schedule, everyone will invest in the signal in exactly the same way, so that they cannot be distinguished on the basis of the signal. (Spence 1973, 358)

More generally, Bliege Bird and Smith (2005, 224) discuss four requirements for costly signaling strategy to remain stable in a population:

1. Members of a group vary in an underlying attribute that is difficult to observe, such as reproductive fitness, wealth and means, job ability, or cooperativity.
2. Observers benefit from accurate information on the level of this attribute; females find more fit mates and employers more able workers.
3. The interests of signalers and observers conflict so that the signaler has the incentive to deceive the observer. This applies to less fit males, con artists, and unskilled employees.
4. Signal cost (or benefit to the observer) is dependent on the underlying quality of the signaler. In the above examples, peacock males with brilliant, unwieldy plumes have to be correspondingly better in evading predators and searching for subsistence; conspicuous consumption of luxury goods cannot be maintained unless the person is endowed with considerable means; university education cannot be attained without several abilities that are in demand. But what signal could be tied to willingness to cooperate?

A recurring suggestion (the "green beard" hypothesis) is that cooperators could identify each other by some phenotypical signal and restrict cooperation accordingly. This is unviable, as there is no reason why evolution couldn't produce an opportunist that also carries this sign, especially since it is unlikely that a single gene would code for both it and the desired behavior (McElreath and Boyd 2007, chapter 5.3). The difference compared to the peacock and education examples is that it is difficult to imagine what marker could have its costs correlated with an individual's willingness to help others.

More viable arguments exist, contending that cooperation itself can be a costly signal of skill, means, and cooperative inclination. Behaving prosocially builds up reputation as a cooperator, which brings future returns.[9] The model by Gintis, Smith, and Bowles (2001) shows that it is a stable strategy for high quality individuals to send costly signals and for everyone to ally with signalers.

9 For example, the Chaldeans take advantage of high broadcast value situations where costly cooperative acts boost their reputations as givers, by arguing over who gets to pay for everybody's dinner in a restaurant (Henrich and Henrich 2007, 63).

The benefit of the signal can be zero or even negative, as long as this lack is compensated by the number and quality of partnerships formed.[10] Group-beneficent signals, such as punishing non-cooperators, will of course increase average fitness and set the stage for cultural group selection. Bowles and Gintis (2003, 437) suggest that signaling and cultural group selection may be mutually reinforcing phenomena, as group selection proliferates cooperative signaling, while signaling creates stable equilibria situations within groups (as established by the model).

What emerges from these various considerations is a multifaceted picture of human cooperation. On the populational level, various mechanisms of stabilizing cooperation exist. But they rely on specific situational constraints in order to adequately distinguish those who reciprocate in kind from those who do not. There is no general-purpose strategy one could adopt and apply successfully regardless of the situation. Henrich and Henrich (2007, 61) note that "strategies that always help (regardless of reputation – unconditional altruists) are highly destructive to cooperation." The ecology of other strategies, possible interpretations of situational cues, and variable group structures create a combinatorial explosion for which there is no catch-all solution – unless one considers the human capacity for cultural learning to be one. Although large-scale cooperation requires considerable mental capacity, especially in the area of social cognition, and is not as such found in other species, one should not equate cooperativity with "intelligence." Henrich and Henrich (2007, 41) contend that evidence shows more intelligence to lead to less cooperation in humans: "In a complicated world with imperfect information, the skills of deception, deceit, trickery, and manipulation, which are improved by some kinds of intelligence, are more effective at destroying cooperation than capacities for tracking past interactions and planning are for sustaining it." Also, other primates are surprisingly ineffective in learning to cooperate despite their cognitive endowment, while communal hunters such as spotted hyenas solve cooperative laboratory tasks with ease, even with little prior training (Drea and Carter 2009). This cross-species evidence

10 The data that best fit this conception (Bliege Bird, Smith, and Bird 2001) takes the size of the catch among spearfishing Meriam of Torres Strait to be a reliable indicator of hunting skills (implying high physical, cognitive, and leadership skills), the presence of which is enough to make the individual a more desirable marriage or political partner – not his proclivity to cooperation. Spearfishing was not in itself as productive as collecting shellfish, and the hunters retained none of the quarry from turtle hunts, so the enthusiasm men showed towards these activities is best explained as a form of costly signaling. Although the authors failed to demonstrate a clear link between signals and subsequent gains from the community, the best hunters clearly had high social status.

suggests that cooperation requires innate dispositions specific to it. To address what these could be in the human case, I will turn next to the sociopsychological capabilities that enable us to observe and follow social norms.

2.3 Social Psychology of Norms

In the introductory chapter (1.2.4) I outlined how adaptive behavior depends on being able to predict the workings of the surrounding environment, and that the existence of other social agents introduces a whole new level of complications. Thus, people act on their expectations of other people's actions and cooperate conditionally based on whether the others will reciprocate (Dawes 1980; Gächter 2006; Yamagishi, Jin, and Kiyonari 1999). But how do they form these expectations? In the previous subchapter I argued that much of our prosocial behavior depends on culturally learned norms of correct behavior. This is not new, as the concept of norm is widely applied in explaining social phenomena. Definitions usually include prescribing behavior that is considered desirable in a given society, and that this desire is to a significant extent shared by its members. And to be a real norm, it is thought that breaking it should provoke at least informal sanctions. But the problem with norms has been that their power seems to be indiscriminate; people occassionally abide by them, but at other times ignore them.

Bicchieri (2006, 11) argues that we should not expect individuals to be unconditionally committed to any norm. Rather, social norms are to be viewed as behavioral rules conditional upon expectations we have of the actions of others. I prefer to abide by a norm I know to exist if: 1) I have categorized the current situation to be of a type this norm applies to, 2) I believe that a sufficient number of others will abide by it (an empirical expectation), and 3) I believe that a sufficient number of others expect me to conform to it and may sanction me if I don't (a normative expectation).[11] If conditional cooperation is deemed socially more acceptable than defecting on a cooperator (which might be penalized with a bad reputation if nothing else), social norms can transform social dilemmas into coordination games (Bicchieri 2006, 2–3, 26–28). In a prisoner's dilemma situation, mutual cooperation and mutual defection become equilibrium posi-

11 Bicchieri (2006, 29–42) differentiates social norms from conventions, fads, and other behavioral regularity, which she groups under "descriptive norms" and "conventions." These lack the third, normative part in the definition of social norms, and work to coordinate action in situations were there is no dilemma involved. When I use the term "norm" I am invariably talking about social norms as defined by her.

tions, as both are more desirable to both players than other alternatives. However, this requires both players to think it likely that they are paired with an individual who follows the same relevant norms, as dictated by the empirical expectation clause above. Bicchieri (2006, chapter 5) maintains that a norm may continue to function despite its being almost universally disliked and the pressure to conform being more imagined than real. As norms are not always discussed openly, individuals engage in social comparison in order to deduce appropriate behavioral rules. As noted above (chapter 2.2.2), punishment or the threat of it is a viable way of stabilizing any norm, and combined with social comparison behavior, witnessing even occasional acts of punishment should be enough to convince uncertain individuals that "others" endorse a norm, easing the task of punishers. Henrich and Henrich (2007, 140) interpret the experimental work of Fehr and Gächter (2002) to demonstrate that "people have both a taste for punishment and an expectation that punishers are out there."

Ylikoski (2011, 52) aptly characterizes norms in Bicchieri's theory as self-fulfilling prophecies. But for a norm to be actually effective, it seems to be crucial that it is made salient, that we are reminded of its existence and applicability in the present context (Kallgren, Reno, and Cialdini 2000). Often a situation can be interpreted in various ways; for example, whether being presented with a gift signifies a reward for a task completed, something that should be compensated later, or a condescending act of charity.[12] Bicchieri (2006, 63–80, 131–133) argues that it would be erroneous to assume that people have internalized norms as generic prosocial imperatives such as always "reciprocate" or "cooperate," as the meaning of these things is very sensitive to context. Cultural expectations acting on situational cues play a large part here, and the interpretation is of consequence to which norms we consider to be applicable. For example, when people consider whether they are being treated fairly, they form expectations based on what others similiarly situated have received. In experimental situations, introducing even a slight cue of one party being more deserving than others results in subjects accepting unequal divisions of the cake, while convincing others there is no alternative makes them accept what would otherwise seem unfair. This is why enforcing a specific reading of a situation is often at the center of

12 In interpreting behavioral game situations, the Detroit Chaldeans were cued by culturally learned norms (Henrich and Henrich 2007, 144–146, 169). In an Ultimatum Game (where it is possible to punish a low offer by rejecting it and preventing both players from receiving anything) they were as fiercely competitive as their business practices are. But removing the rejection option (which turned the experiment into a Dictator Game) brought about a reinterpretation of the game as acts of charity, with little feeling of competition.

disputes, as ambiguity invites self-serving interpretations.[13] As Bicchieri (2006, 90–98) suggests, norms are best viewed as being embedded into cognitive scripts, activated once a situation has been categorized, and providing participants with roles that predict their behavior. This categorization process is dependent on comparison with past experiences, and it is quite possible that participants arrive at different conclusions as to its nature, and may activate different scripts. This may result in breakdown or renegotiation of appropriate behavior. Norm activation is thus a question of cognitive heuristics, and it is easy to interpret various cultural conventions as geared towards diminishing ambiguity as to the nature of a situation or its participants, such as ethnic markers (Gil-White 2001) and ritualized pledges (Rappaport 1979a; see chapter 1.2.3). The Detroit Chaldeans, for one, use both language and religion to index adherence to shared social norms, as those who had dropped out of the Chaldean church were seen as "fakers" who had also lost their norms, and recent immigrants from Iraq struggle to learn the language that is much more important in the Detroit context than in their homeland (Henrich and Henrich 2007, 178, 181–184). According to Atran (2010, 431), studies show that people seldom bother to even check the reputation of outgroupers, since they are assumed to have incompatible goals.

That humans are prone to favor in-group and discriminate against out-group even in situations where these groupings are based on arbitrary differences was established by the minimal group experiments of Tajfel (1970). This result surfaces recurrently. Bicchieri (2006, 153–154) describes an experiment where many subject groups concluded that the members of out-groups are likely to defect on them, as this was what they would do themselves. They considered such discrimination to be normal behavior, not resulting from antisocial dispositions of the out-groupers. Social psychologists have speculated whether this results from a generic norm of in-group favoritism, maintaining positive group identity, or something else. However, analyzing several experiments, Yamagishi, Jin, and Kiyonari (1999, 181–186) concluded that subjects in a minimal group experiment do not favor in-group members unless they have reason to expect reciprocal fa-

---

13 The poor view of the power of norms in enforcing behavior may result from what has been (somewhat overstatedly) called "the fundamental attribution error" (or "correspondence bias," Gilbert and Malone 1995) that is common among Western populations: to interpret the behavior of others as resulting not from peer pressure or situational role but from private preference, while there is much evidence that the most important determinant of behavior is the current situation or one's reading of it (e.g., Latané and Darley 1968; Brewer and Kramer 1986; Kallgren, Reno, and Cialdini 2000). East Asians, for one, are much more likely to endorse contextual explanations for others' behavior (Henrich, Heine, and Norenzayan 2010).

voritism from them. For example, if in a prisoner's dilemma experiment the partner does not know he has been paired with an in-group member, the subject (being aware of his ignorance) does not favor him over a partner from an out-group. When favoritism occurs in conditions of symmetrical information, there is an implicit expectation of reciprocation present. Again, prosocial action is conditional on expectations. Yamagishi, Jin, and Kiyonari derive from this a theory of "bounded generalised reciprocity," where a favor performed to an in-group member is expected to be repaid not necessarily by the same individual but possibly by another in-grouper (compare with "indirect reciprocity" in chapter 2.2.1). Post-experiment questionnaires reveal this expectation to have been at work in minimal group experiments. It is important to note that this was not illusory – group favoritism happens, and it pays to be part of such a system. But when we move to less clearly delineated, real world groups, the crucial question is, how do these groups keep track of their membership and ensure that those accepted are going to grant the favors expected? Casting the problem of reciprocal cooperation in terms of expectations and the expectations of others' expectations – that is, social norms as defined above – provides a more psychological-level account of how cooperation can work but does not in itself resolve the key issue of recognizing commitment.

2.4 The Strategic Role of Emotions

According to Damasio (2004), emotions consist of a physiological process that leads from appraisal of an emotionally-competent stimulus to an emotional state that is felt both in the body proper and in the central nervous system's representation of it. This is to be distinguished from a state of feeling which is the joint perception of the causative stimulus, the emotional state, and the thoughts and cognitive states it may trigger. Put another way, emotions have a regulatory function by enjoining internal and external information to the individual into a form that is conducive to decision making, thus working against the more negative forces that threaten the individual and are as such important for survival and well-being. For example, encountering an angry dog may evoke a feeling of fear that effectively blocks approach behaviors (Damasio 1996).

Classic results in the psychology of emotions indicate that emotions consist of two factors (Schachter and Singer 1962; Dutton and Aron 1974): a physiological reaction invoked by a situation and a cognitive labeling based on contextual cues. Mainly we think of emotions in terms of such linguistic labels – being happy, afraid, envious, and so forth. The case that some of these folk categories correspond to universal basic emotions has most prominently been made by

Ekman (1994, 271, 278–281), who originally listed anger, disgust, fear, happiness, sadness, and surprise as being cross-culturally recognizable from facial expressions (although he has later substantially expanded this list; see Ekman 1999). This has led to an ongoing debate, as some argue that cultural learning can customize the expression and recognition of these emotions (see Prinz 2004, 78).

The problem with (social) constructionist alternatives to the basic emotions theory has been that they seldom adequately address evolutionary or neurocognitive questions. A significant new opening is the conceptual act theory of emotion (Wilson-Mendenhall et al. 2011; Lindquist et al. 2012, 123–126), based on the idea of *psychological* construction of emotions in the brain. Physiological core affect is conceptualized in a process which involves comparison of the current situation with prior experience stored in concepts (loose aggregates of situated exemplars and appropriate responses). It is this categorization which anchors the experience to stored emotion words. The process also involves executive attention at different stages to determine the most meaningful conceptualization and response. On the brain level, this is realized as assemblies of neurons within distributed networks combining and constraining one another in real time.

With recent advances in neuroimaging techniques, the comparison of these models can be taken to the neuropsychological level. Vytal and Hamann (2010) concluded in a meta-analysis of 30 studies that five of the Ekman's original basic emotions have consistent and discriminable neural correlates (surprise being difficult to study in a controlled way). In contrast to this, Lindquist et al. (2012) performed a similiar meta-analysis, finding support for their psychological constructionist approach. One reason for this opposing result lies in the slightly different but equally acceptable approaches taken by the authors of the two studies to comparing activations produced by different emotional stimuli (pairwise contrasts vs. contrast against average), which limits the power of meta-analyses to solve the issue (Hamann 2012). However, Lindquist et al. somewhat misrepresent the basic emotions model as requiring that for each emotion invoked there is consistent and discriminating activation only in a single brain area. Instead, the Vytal and Hamann study presented evidence of discrimination by joint activation, which is in keeping with the current, network-oriented view of brain function.[14] However, Hamann (2012, 460–462) has concluded that the evidence

---

14 Lindquist et al. (2012, 129–136) refer to conventional views which locate fear in the amygdala, disgust in the anterior insula, anger in the orbitofrontal cortex, and sadness in the pregenual/subgenual anterior cingulate cortex. In the Vytal and Hamann (2010, 2879) study, fear consistently activated both amygdala and insula; disgust both inferior frontal gyrus and anterior insula; anger both inferior frontal gyrus and parahippocampal gyrus; sadness both middle

for neural correlates of basic emotions is currently inconclusive and that the network models require re-evaluating the criteria for basic emotions. In future, multivoxel pattern analysis may open new avenues for this, as it has made it possible to identify supramodal, emotion-specific activity patterns on an even finer-grained level (Peelen, Atkinson, and Vuilleumier 2010). Also, from the functional network perspective the two models may complement rather than contradict one another.

Basing their approach on the basic emotions theory, several authors have dissected the ways in which specific emotions appear to have been shaped by natural selection during human evolution to perform regulatory tasks within or between individuals. Tooby and Cosmides (2008) argue the function of emotions to be coordination among competing subprograms of the mind. For example, guilt would work to recalibrate variables controlling tradeoff between self and another in cooperative relationships in order to preserve them. Fehr and Gächter (2002) report that the proximate factor to triggering punishment is the emotion of anger, which scales with the perceived difference between one's own commitment and that of the defector, and that defectors expect this to be so. Shame and pride are connected to norm obedience and advertizing one's attractiveness as a cooperative partner, and mirth builds solidarity, as work groups that laugh together work more efficiently (for more examples, see Fessler and Haley 2003, 14–27).

While these functional couplings are plausible, these accounts do not detail how they assume these emotions are realized in the brain. As Lindquist et al. (2012, 141) note, in basic emotions accounts an emotion is often equated with a specific behavior and its neural circuitry, which limits the applicability of a complex psychological category. There are many situations that can elicit the emotion of fear (encountering a dark alleyway, memory of a catastrophe, choosing a secure password) and many adaptive responses to it (escape, attack, deliberation). Similiarly, there are many ways variables can be recalibrated in a relationship (being frank or telling lies, redefining the relations, etc.), and the emotional experience and associated neural processes vary accordingly. Even if some brain areas are consistently activated for a basic emotion, making sense of how this applies to the situation requires considerable psychological construction (Wilson-Mendenhall et al. 2011, 1108).

Recognizing emotions is an issue somewhat distinct from experiencing them: the amygdala, for example, is activated only by the perception of fear

frontal gyrus and head of caudate/subgenual anterior cingulate cortex; and happiness both rostral anterior cingulate cortex and right superior temporal gyrus.

(alongside other ambiguous but motivationally salient sensations), not by experiencing it (Wilson-Mendenhall et al. 2011, 1121). But there is also evidence that recognizing an emotion in another produces synchronous brain activity across individuals, which could mean that emotions promote social interaction by simulating others' mental states (Nummenmaa et al. 2012). This fits well with the influential hypothesis of Frank (1993) that emotional expression has the social function of signaling commitment. In some situations it is materially advantageous for an individual to be known as honest or vengeful to the point of disregarding material gains; for example, the relentless pursuit of petty thefts. While emotions give individuals the impetus for this kind of behavior, the strategic effect is achieved by reliably communicating these commitments, either by spectacular past actions or difficult-to-fake bodily signals. Basic emotion theorists have argued that both suppressing emotions and producing them voluntarily is difficult, as coordination of face muscles required for smiling is beyond conscious control, as is the prevention of the "micro expressions" they generate before being contained (Ekman, Friesen, and O'Sullivan 1988). But Barrett, Mesquita, and Gendron (2011) have also demonstrated that contextual information (surrounding stimuli, perceiver state, cultural biases) have an early and automatic effect on emotion perception. Taking this into account rather strengthens the case that as emotions are not easily faked, they cause observers to assess them as reliable indicators of beliefs and motivations held by an individual. This enables emotions to elicit desirable responses in others, turning them into strategic devices; once the feeling is present, others are drawn to it (Bulbulia 2012; cf. Nummenmaa et al. 2012).[15]

An important related issue is that emotions can also be experienced in a corporate fashion, triggered by actions that affect a group that the subject feels part of (Fessler and Haley 2003, 23, 26–27). In such cases, the fortunes of the individual are often aligned with the fortunes of the group, and the display of emotions signals this alignment to others. For example, moral outrage, an emotional state resulting from norm violations being experienced as transgressions against one-

15 Surprisingly, there is evidence that individuals are also sensitive to cues that predict this strategic value in a given situation (Griffits 2004, 101). For example, athletes and their audiences expressed smiles over success mostly when facing a social interactant such as another audience (Ruiz-Belda et al. 2003). Also, there is currently nondecisive evidence on whether untrained individuals can discern honesty from nonverbal cues such as facial expressions. Studies have found moderate evidence for university students' ability to recognize cooperators and defectors (Brown, Palameta, and Moore 2003; Yamagishi et al. 2003; Oda et al. 2009a, 2009b). Some groups of experienced forensic professionals show remarkable skill in detecting liars (Ekman, O'Sullivan, and Frank 1999), and this may be because they pay attention to the correct signs: micro-expressions of fear, excitement, and guilt (Vrij et al. 2000, 241–242).

self, makes the punishment of norm violators spontaneous. Also, Haidt (2003) attempts to distinguish a class of "moral emotions" that are elicited by events not directly related to the self but which still motivate prosocial action. Of these, anger, contempt, and disgust would cause people to alter their relationships to moral violators, while self-conscious emotions (shame, embarrasment, and guilt) should elicit conformance to rules and social order.

2.5 Chapter Conclusions

In this chapter I have described why human ability in solving social dilemmas continues to pose a problem to evolutionary and behavioral sciences. Of various evolutionary solutions proposed, no single mechanism suffices to explain the scope of human prosociality. The cultural learning and group selection theory points towards an answer that would account for the variability and historical development of cooperative societies, but one which requires more detailed histories of how various cultural groups have "locked" into their set of norms and thus gained an upper (or lower) hand *vis-à-vis* some other group. The present study should be seen as a part of this endeavor.

On the behavioral level, Bicchieri's (2006) argument concerning social norms provides several convincing points regarding how individuals observe, appraise, and conform to norms in their environment. It also has the attractive quality of according well with current theories about our cognitive abilities and the evolution of cooperation. For humans, norms constitute a crucial part of the strategic environment, as they affect the perceived payoffs in social dilemmas. What arises as the key factor is the expectations a person has: of the applicability of a norm in the given context, of others' willingness to observe and uphold the norm, and of who constitutes the "others" from whom one can expect reciprocation or punishment. For this reason increased ambiguity and uncertainty in a social environment should be associated with less norm-compliance, unless cultural conventions that diminish these are employed.

But emotions also have properties that appear to solve the problem of commitment to a group or a course of action. They function as proximate factors that help evaluate between behavioral outcomes and trigger a particular response. This involves involuntary corporal reactions that are dependent on past experience and a reading of the situation. For this reason, observers (especially coethnics and others from "psychologically similiar" cultures; see Barrett, Mesquita, and Gendron 2011) can also use them as reliable indicators of how the individual is inclined to behave in a similiar situation in the future. Deducing the beliefs and behavioral dispositions another person is committed to constitutes strategic information.

What emerges from these considerations is that in naturalistic settings individuals can rely on a combination of heuristics – ethnic markers, emotional expression, accompanying behaviors, contextual cues – to decide with whom to affiliate even in the absence of first-hand experience or reputational information. "Mormons are no doubt better than non-Mormons both at recognizing Mormons and at recognizing non-Mormons," (Bacharach and Gambetta 2001, 171) since to reliably pose as a member of an idiosyncratic group one has to concurrently master several signs that come naturally to a true member, the absence of which is striking to those used to them. With emotions, this may even be impossible, as "absence of fear as the lions charge the Coliseum floor is possible only if one believes in another world that will not end in [the lion's] jaw" (Bulbulia 2012). Starting with Irons' (2001, 298; 1996) contention that "religious traditions are sufficiently complex to be difficult for an outsider – one who has not spent years learning the tradition – to imitate," there has recently been considerable interest in the way shared religion appears to solve social dilemmas and enable cooperation on a larger scale. In the next chapter I will review the empirical evidence and discuss theoretical explanations that have been offered to account for these phenomena.

3 Religion as a Solution to Social Dilemmas

> Only by God's help in establishing His religion do individual desires come together in agreement to press their claims, and hearts become united. The secret of this is that when hearts succumb to false desires and are inclined toward the world, mutual jealousy and widespread differences arise. When they are turned toward the truth and reject the world and whatever is false, and advance toward God, they become one in their outlook. Jealousy disappears. Mutual co-operation and support flourish. (Ibn Khaldûn 1967, 125–126)

As noted in chapter 1.2.3, the suggestion that religion brings social cohesion dates back to Durkheim (1915) but has lacked a plausible theoretical explanation of why this should be so. The association of religion to morality or moral behavior is even more common, but there are varying interpretations as to what kind of causal link there is, if any (Pyysiäinen and Hauser 2010, 105–106). For example, the classic "Good Samaritan" experiment failed to demonstrate an association between having one's mind occupied with the biblical parable of altruism and actually performing an altruistic act (Darley and Batson 1973). But the problem with such approaches may have been that they have approached religions as providing for general-purpose, context-free altruism. As argued in chapter 2.2, non-discriminatory cooperators face extinction and may even be destructive to cooperation. In a study conducted in two American communities (Orbell et al. 1992), significant cooperation between strangers was found only for Mormons in Logan, Utah, where 75 percent of the population consisted of their coreligionists (they had a far above average chance for a "stranger" to be another Mormon). Similarly, in sociological studies (reviewed by Stark 1996) that examined the connection between individual religiosity and delinquency in America, geography arose as the determining factor: only West Coast data yield no relationship, and Stark explains this as resulting from the fact that church membership rates there are among America's lowest, as religion empowers norm conformity "only as it is sustained through interaction and is accepted by the majority as a valid basis for action."

This accords well with the account of norms given above, and in this vein, a review of existing research by Norenzayan and Shariff (2008, 58) argued that the effect religion has on prosocial tendencies would be context-sensitive with clear boundary conditions and depend on reputational sensitivity, a general psychological mechanism found in humans: "The cognitive awareness of gods is likely to heighten prosocial reputational concerns among believers, just as the cognitive awareness of human watchers does among believers and nonbelievers alike." The difference is that if one genuinely believes in supernatural monitoring, one believes it is omnipresent. These considerations make for predictions which I will review next, alongside empirical evidence, some of which was not cited by Norenzayan and Shariff.

3.1 Effects of Religion on Prosocial Behavior: Four Predictions

3.1.1 Reputational Concern

Religious devotion is expected to be associated with greater prosocial reputational concern. Those who engage in religious activity report more charitable behavior than those who do not, a convergent finding of many studies that holds even after controlling for various other variables (see Monsma 2007 for a review of American surveys). But Norenzayan and Shariff (2008) argue that as this evidence is based on self-report, it is suspect to "impression management" (Paulhus 1984, 604–605), a human tendency to present a more positive view of the self when questioned. They cite the meta-analysis of Trimble (1997) to demonstrate that measures of religiosity are positively associated with a tendency to project an overly positive image of oneself in evaluative contexts, and the Batson, Schoenrade, and Ventis (1993, chapter 10) review of experiments in which even the Allport and Ross (1967) measure of intrinsic religious orientation is not so much correlated with sincere concern for other's welfare as with a desire to appear concerned for purposes of impression management. Extrinsic religiosity predicted helping only when not helping would make one look bad, and in their studies only the "quest" dimension of religiosity (Batson, Schoenrade, and Ventis 1993, chapter 6) was related to increased sensitivity to the needs of others.

What Norenzayan and Shariff do not mention is that while the review by Trimble (1997, 980–983) does report a positive relationship between measures of social desirability and intrinsic religious orientation, there is a lack of relationship between this desirability and extrinsic orientation. He also notes that it may not be valid to assume that reports of "righteous behaviors" result from socially desirable response bias rather than actual behavior. Questions used for measuring social desirability are to an extent congruent with those measuring intrinsic religiosity (e.g., "I always try to practice what I preach"), the underlying assumption being that no one actually behaves in such a "pious" manner, and Trimble suspects this may not hold with highly devout people such as those who score high on intrinsic religiosity. Of course, every research method has its potential artifacts, but the robustness of a finding is determined by whether it recurs across different operationalizations operations and methods, and apart from those cited above, such evidence of religion heightening reputational sensitivity is hard to find.

Furthermore, all of the studies reviewed by Batson, Schoenrade, and Ventis (1993, chapter 10) had American undergraduates as participants, and presumably this applies also to those in Trimble (1997). It is noteworthy that in the latter case, the "conspicuous outlier" was a study by Hunsberger and Ennis (1982) where Catholic students displayed *less* socially desirable responding when the experimenter was dressed as a priest. Considering the very justified concern over representability of American undergraduates (Henrich, Heine, and Norenzayan 2010, 5–7 present some evidence of cultural variability in reputational sensitivity), these data do not yet warrant generalization over all cultures as to whether religiosity is associated with an increased reputational concern.

3.1.2 Cues of the Supernatural

Cues of the supernatural create a sense of being monitored. Boyer (2001, 150–160) characterizes gods and spirits as agents that have full access to strategic information. Consequently, attending surroundings containing cues of the supernatural (as is common in ritual performances) should activate thoughts of morally concerned deities, if appropriate, and facilitate prosocial behavior. In chapter 2.3 it was argued that for a norm to be effective, it has to be "primed" so that we remember its existence and possible applicability in the present situation. A clear connection between religiosity and cooperation was established in economic games organized in secular and religious Israeli *kibbutzim* (Sosis and Ruffle 2003, 2004; Ruffle and Sosis 2006; more on these in section 3.1.4). As Norenzayan and Shariff note, the latter is likely to have reminders of the supernatural in abundance, and in these cases the benefits of prosociality were clearly confined to ingroup members.

But to approach the question of omnipresent monitoring from another angle, a growing body of evidence shows that artificially inducing thoughts of being watched increases cooperative behavior. American students who heard a story of a ghost being spotted in the room cheated less on a computer task (Bering, McLeod, and Shackelford 2005), and children who were told of an invisible watcher ("Princess Alice") refrained from peeking inside a forbidden box (Bering 2004, 425–426). In the case of the pair of eyes placed above the honesty box for payments for tea, coffee, and milk in a naturalistic workplace environment, as discussed above, payment increased dramatically (Bateson, Nettle, and Gilbert 2006). Both Haley and Fessler (2005) and Burnham and Hare (2007) received similiar results in economic games where experimental manipulation involved a desktop background picture with a pair of eyes, leading to an increase in prosocial behavior. Verbal cues of monitoring also appear to work, as subjects who unscrambled (dropped off one and arranged the rest in a meaningful order) unintelligible sentences containing words relating to religion ("prophet," "sacred") or secular jurisdiction ("police," "contract") behaved more prosocially in a dictator game (Shariff and Norenzayan 2007; Randolph-Seng and Nielsen 2007 received similiar results).

3.1.3 Morally Concerned Deities

Morally concerned deities should enable a larger society size. As described above (chapter 2.2.1), an increase in group size has an inhibiting effect on cooperation, as larger groups are more vulnerable to freeloading and subsequent collapse, which prompts a need for institutional supports. Norenzayan and Shariff (2008) predict that large societies which have stabilized high levels of cooperative norms are more likely than smaller ones to espouse belief in deities that are actively concerned with human morality. Such "supernatural punishment hypothesis" has also been laid out in detail by Johnson (Johnson and Krüger 2004; Johnson 2005; Johnson and Bering 2009). This line of inquiry suggests the evolution of religions via cultural group selection, as selection pressures created by the growth of agricultural societies into larger and more complex ones would have favored groups with moralizing deities (Shariff, Norenzayan, and Henrich 2009).

Using *Ethnographical Atlas* (EA, 1267 societies) and the more representative *Standard Cross-Cultural Sample* (SCCS, 186 societies), Roes and Raymond (2003) found a modest but highly significant correlation between society size and likelihood of belief in morally concerned gods. Also using SCCS, Johnson (2005) found belief in such "high gods" to correlate significantly and positively with cooperativity-implying variables.[1] Finally, Snarey (1996) found in SCCS a strong and highly significant association between moralizing gods and scarcity of water, a resource critical to group survival.

1 These were society size, trust-based monetary systems, centralized enforcement and sanctioning systems, and viable taxation systems. Norm compliance was significantly correlated in some tests, but not in others, alongside community loyality, food sharing, and lack of internal conflicts. One of five measures of internal warfare frequency was even positively correlated with high gods, contrary to predictions (increase in factionality is a possible reason for this; Johnson 2005).

The statistical tests reported above operate on a very high level and conceal substantial unit level variation that can result from numerous sources other than the hypothesized fear of supernatural punishment. Nor is the variable measuring belief in a high god, i.e., "a spiritual being who is believed to have created all reality and/or to be its ultimate governor, even though his sole act was to create other spirits who, in turn, created or control the natural world," a perfect index of expected divine retribution. However, the higher levels of coding explicitly mention the deity as being interested in human affairs (level 3) and supportive of human morality (level 4), as world religions vary in regard to specifically what transgressions the high gods (and not some other entitites) are expected to punish (Johnson 2005, 418–420). But the noise in the data contributed by these sources of error is more likely to work against the hypothesis (increase the likelihood of type II error), making the analysis conservative unless there is some alternative source of punishment that covaries with high gods. But two experimental studies point away from this option. First, in the Henrich et al. (2010) study mentioned above, the single experiment in which association between world religion and fairness was *not found* was a third-party punishment game, in which a third player is given the possibility to punish low dictator game offers; the covariance was thus inverse. And second, in a mathematics task which allowed for cheating, belief in a punitive god was associated with less cheating, while belief in god in general was unrelated to it (Shariff and Norenzayan 2011).

3.1.4 Signals of Commitment

Religious behavior that signals commitment should mobilize greater cooperation and trust. This should be especially true if reputational information is not available. In a laboratory experiment with German students, those deemed particularly "religious" (Judeo-Christian) according to a questionnaire exhibited more trust towards persons they were told were also religious, and the religious individuals were also more likely to reciprocate this trust (Tan and Vogel 2008). Also, *madrasah* imam pupils behaved more prosocially towards one other than college students in an experiment organized in an Indian rural community (Ahmed 2009); their identities were kept secret except for the fact that other participants were from the same school. Iannaccone (1994) found from American survey data that the stricter the denomination (Christian or Jewish), the more members contribute in terms of time and money, and the less they participate in outgroup activities. Also, a survey reports that compared to religious and nonreligious groups, Americans perceive atheists as the least trustworthy group (Edgell, Gerteis, and Hartmann 2006).

Ethnohistorical studies on how the spread of Islam may have facilitated long-distance trade in Africa (Ensminger 1997) and how medieval Maghrebi trade organized around Jewish peer communities (Greif 1989, 1993) are compatible with hypotheses that predict greater intragroup trust among those with simiiliar religious commitments due to lower monitoring costs (see North 1985 on the importance of transaction costs). Experimentally, Henrich et al. (2010) show that over 15 populations (only one Western), both market integration and participation in a world religion (Islam or Christianity) are associated with fairness in economic games, although the latter effect was significant in only two out of three experiments (more on this below). An experiment by Bulbulia and Mahoney (2008) also demonstrated that a costly parochial act committed in the name of Christianity can motivate supranational, non-reciprocal cooperation, whereas one committed in the name of nationalism does not have the same effect on one's fellow citizens.

But the commitment hypothesis does not apply only to world religions; Soler (2008) conducted a field experiment among a Brazilian Candomblé community (a religion that resulted from the nineteenth century mixing of an indigenous Brazilian belief system and the beliefs of enslaved West-Africans with some Christian elements), and found cooperation in an economic game to correlate with a group commitment subscale factored out of a religious commitment scale, but not with a personal commitment factor. Similiarly, that ritual participation, and not belief or narratives alone, fosters support for suicide attacks (an extreme form of group commitment) is suggested by Ginges, Hansen, and Norenzayan (2009), who conducted four separate studies: two surveys of Palestinian Muslims; a priming experiment with Israeli Jews; and a cross-cultural survey of Indonesian Muslims, Mexican Catholics, British Protestants, Russian Orthodox in Russia, Israeli Jews, and Indian Hindus.[2] In the Israeli experiment a prayer prime resulted in slightly lower support than a no-prime condition, a finding consistent with Lambert et al. (2010), who found that prayer for a partner increased forgiveness.

Pertaining to the signals of commitment prediction, a series of studies conducted by Sosis et al. offer substantial empirical support to a commitment theory of religion based on the concept of "costly" or "hard-to-fake" signals, an idea

2 In the Ginges, Hansen, and Norenzayan (2009) study, respondents were asked whether they would be willing to die for their god/beliefs and whether they blame people of other religions for the troubles of the world.

first advanced by Irons (1996). As described in chapter 2.4, reliable signals of commitment manipulate audiences, and *ceteris paribus*, "the costlier the signal, the less likely it is to be false" (Irons 2001, 298; or there is an optimal level of costliness, see Sosis and Bressler 2003, 234). Most religious traditions involve substantial sacrifices and entail loyalty to a particular community and its norms. But the mutability of cultural signals also enables us to flexibly alter our loyalties, adjusting to new situations as they arise. Sosis, Kress, and Boster (2007) compared male ritual marking across 60 temporally and geographically distributed cultural groups using Human Relations Area Files. They found that societies that frequently war against external enemies apply permanent markers (piercings, scars, mutilations) to signal group identity, whereas those where internal warfare is more prevalent (today's enemies may be tomorrow's allies), engage in ritual activities that do not leave a permanent sign (body painting, toxic substance ingestion, isolation) and thus allow individuals to flexibly switch alliances.[3]

The fact that cooperation is required more urgently in some times and places than in others, and that religious groups come in many varities, from small communities to universal faiths, guarantees a complex dynamic of varying signal costs and mediums. This can be seen most clearly in the case of closed communities, whose life is rife with social dilemmas: mindful consumption of common food and other resources, temptation to shirk from working but partake in the benefits, and suspending independence in favor of group interests (Kanter 1972, 65–74; Taylor 1982, 120–125). All these factors contribute to the risk of the commune dissolving. Sosis (2000) found that, of nineteenth century American utopian communes, the mean lifespan for secular communities was 6.4 years and for religious communities 25.3 years. Further analysis by Sosis and Bressler (2003) assessed the amount of costly, hard-to-fake requirements made by each commune, which revealed that the duration of religious communes correlated with the number of costly requirements they imposed, but this did not hold for secular communes.[4] Sosis and Ruffle (2003; also Sosis and Ruffle 2004; Ruf-

3 Sosis, Kress, and Boster (2007, 245) note that this finding supports the converse of a frequently espoused view that (extreme) religious behavior causes warfare. The theoretically more plausible explanation here is that prevalence of warfare leads to ritual extremisms. However, they note that the relationship may be mutually reinforcing, as the ritually reinforced intra-group trust enables a group to also mobilize for acts of war.

4 The original dataset in Sosis (2000) included 200 communes, and Sosis and Bressler (2003) could determine costly requirements for 83 of them. Seculars include socialist, anarchist, Owenite, and Fourierist communes; Hutterites were excluded from the religious due to their extraordinary success, which would have biased the results. Costly requirements were dete-

fle and Sosis 2006) obtained comparative results from contemporary Israeli *kibbutzim*; the secular *kibbutzim* are rooted in socialist Zionism, while the religious kibbutzniks follow "Modern Orthodoxy" Judaism and adhere to *halacha*, the traditional Jewish law. Ritual plays a large part in their lives, with men having publicly oriented ritual requirements (prayer in synagogue), while women pursue theirs privately at home. In an economic game that measures trust between two participants,[5] males from religious *kibbutzim* who attend synagogue daily were significantly more trusting than females from the same kibbutz or members of secular *kibbutzim* (with whom no gender difference existed). This trust correlated with frequency of synagogue attendance.

But it is not only the costliness (or number) of requirements that matters but also the motivational ideology. This is underlined by the above finding that for secular American communes, costly requirements did not predict longevity. Fishman and Goldschmidt (1990) report that the secular *kibbutzim* are staggering under debt and survive through government subsidies and Jewish philantrophy, while the religious *kibbutzim* show outstanding economic performance and have had no need of outside assistance (the data do suggest stronger measures of mutual aid between the religious *kibbutzim*), despite relative handicaps such as ritualistic-legal norms and a lower level of industrialization. In every decade of their 70-year existence, the per capita net production of religious *kibbutzim* has exceeded that of their secular counterparts. Sosis and Ruffle (2003) add to this that while time spent on *kibbutzim* correlated negatively with trust (possibly due to initial enthusiasm waning as the realities of communal life become apparent), this effect was not found in male religious kibbutzniks; ritual attendance appears to counter it. On secular *kibbutzim*, attendance at communal meals had a "marginal effect" on cooperativity, indicating that they serve a weak, quasi-ritual function there. Of American communities, Sosis (2000) found the secular

rmined by counting the number of requirements entailing temporal, energetic, or financial costs which did not directly benefit an individual's fitness and restrictions that limit his or her possibilities to receive these benefits from nongroup members. These included restrictions on foods and beverages, technology and other items, communication with outsiders, sexual and familial relations, and limits on knowledge and mutual criticism (the costliness of some of these practices depends on their limiting interaction with outsiders, who may for example have abhorred non-monogamous marriages).

5 In Sosis and Ruffle (2003), two players were instructed to decide independently how much money to draw from an envelope known to contain 100 shekels. If the sum of the two draws exceeded 100, neither player got anything. If they were less than or equal to 100, both got to keep what they chose to draw, while the remaining money was multiplied by 1.5 and the resulting amount was split evenly between the two participants. The amount of money drawn by an individual was taken to be an inverse measure of trust and cooperativity.

ones to be more likely than the religious ones to dissolve during the early years of their existence (and every year thereafter), indicating that the difference in survival does not result only from members losing faith in secular ideologies more quickly than in religious ones. The secular ones were also more likely to dissolve due to internal disputes or economic failure, regardless of how many constraints they imposed, which leads Sosis and Bressler (2003, 227) to conclude that a theory that focuses solely on the amount of requirements fails "to capture some critical elements of religious belief that distinguish it from belief in a secular ideology."

The samples in studies reported in this subchapter are somewhat biased towards Western, monotheistic traditions (barring the studies of cross-cultural files). Iannaccone (1994, 1183) actually restricts his argument about strictness and commitment to Western religions, "Christianity, Islam, and Judaism [that] place greater emphasis on collective, congregational activity than do Buddhism, Hinduism, or Shinto." Regrettably, none of the studies that sample societies globally test whether these exists such a division within the data (which would presumably dampen effect sizes or create a continuum). However, there are several more informal accounts on Asiatic religions that also stress their ability to solve social dilemmas. In Wilson's (2005) study on 35 randomly drawn religions, 13 were Asiatic (the Middle East excluded), and the features of Jainism that support group action are discussed prominently. Paxson (2004) offers a similiar account on Sikhism, and Iyer (1999, 103) concludes that "[a]mong business in much of Asia, community and religious identity and historical family lineage play a part in providing a 'club'-like status to the trading group such that contractual uncertainties are reduced in dealing with 'club'-members." Also, the review of Henrich, Heine, and Norenzayan (2010) suggested that "Americans of only 50 or 100 years ago were reasoning in ways much more similiar to the rest of the non-Western world than Americans of today" (with the most significant shift being from holistic to analytic reasoning), in light of which Sosis' commune study can be taken to be more representative. However, more research is certainly needed to clarify this purported difference between "Western" and "Eastern" religions.

3.2 Evaluation of the Signaling Hypothesis

The theoretically most advanced hypothesis that has been put forth to account for the above phenomena is the costly signaling theory of religion as formulated by Irons (2001) and Sosis (2003b, 2006). Within it, behavioral requirements that impose a handicap on individual survival or reproductive capacity either directly or by limiting his or her possibilities to cooperate with members of the out-group are seen as screening out those who are not committed to the ideology of the in-group. Such individuals would presumably be more likely to free ride on the achievements of more committed members, and thus undermine the collective effort.

While the theory is in principle plausible, and the data offered to support it quite strong, a closer look reveals theoretical problems that are hard to resolve. First, there appears to be no inherent reason why costly behaviors should be a reliable indicator of underlying commitment. This is a serious problem, and the solutions proposed by Sosis address it only on the proximate level. Second, costly behaviors predict cohesion only when coupled with a religious ideology. For this problem there are plausible answers within the framework proposed by Sosis and others, although these are yet to be integrated into a clearly predictive model. Third, there is confusion concerning the strategic role of religious emotions in relation to costly ritual signals, and the necessity of one or the other. As of yet, none of these problems have been settled, and below I attempt to synthesize some of the proposed solutions.

3.2.1 The Problem of Underlying Quality

Sosis (2003b, 100–101) argues that "[r]itual performance is a signal that advertizes an individual's level of commitment to the group. Preferred signalers are those who are highly committed to the group and are thus likely to be cooperators." But as seen in chapter 2.2.3, for costly signaling to be an evolutionarily stable strategy, certain conditions must apply, one of which is problematic for the religious signaling hypothesis – signal cost must be dependent on the underlying quality (here, commitment or cooperativity), i.e., an identical signal is costlier to send for a low-quality (less committed) individual than for a high-quality one (Bliege Bird and Smith 2005, 224). Sosis has offered three accounts of how this requirement can be met:

1. Those who have internalized the group's ideology perceive the costs to be lower since they are discounted with supranatural rewards such as a pleasant afterlife (Sosis 2003b, 101–105). Ritual activity serves the proximate function of enhancing belief internalization through mechanisms of cognitive dissonance and self-perception.[6] This account is aimed at providing a

6 Sosis (2003b, 97–98) reviews psychological literature that shows behavior can impact attitutes, opinions, and beliefs, reinforcing the tendency to act similiarly in the future (e.g.,

proximate-level solution only of what psychological mechanisms underlie differential commitment to ritual groups.

2. Those who have already undertaken bans that restrict their non-religious activities consequently have reduced opportunity costs for religious activities (Sosis 2006, 65–70). If watching TV is already forbidden, spending one's time studying sacred texts is not such a big sacrifice. Here, Sosis (2006, 73) is also discussing the ultimate-level issues of evolutionary gains in fitness. But the plausibility of this solution depends on there being a stable environment of interlocking norms, the creation of which requires cultural transmission mechanisms outlined by Henrich (2004, 2009).
3. Sosis, Kress, and Boster (2007, 235) note that the behavioral ecological model of signaling is applicable to religious behavior also if the benefits of costly commitment are delayed, in which case those unwilling to commit in the long term would have no incentive to fake the signals. An example of such limiting was seen during the American Second Great Awakening, when some preachers tested the newfound enthusiasm of their converts by putting them on probation of four to six months before acceptance into the church (McLoughlin 1978, 122). But this is unlikely to be a general explanation for the stability of ritual signaling phenomena.

While clarifying the proximate mechanisms that regulate signal costs and their appeal to believers and non-believers, these explanations only partly address the ultimate level question of how ritual signals could have evolved. In evolutionary time, it has not been the perceived but rather the actual earthly costs and benefits of committing that have determined the fittest strategy. If there is no statistically reliable link between signal cost and signaler cooperativity, a population of sincere signalers could be invaded by signalers who do not internalize beliefs or engage in private rituals and defect when it is most opportune (cf. Henrich 2009, 256; Murray and Moore 2009, 234–235).

Henrich also questions how an individual-centered signaling hypothesis could predict the group-level historical dynamic observed by Sosis and Bressler (2003) without being embedded in a cultural evolutionary framework. Instead, Sosis is working from the hypothesis that the capacity for forming religious sys-

---

Freedman and Fraser 1966; DeJong 1979; Burger 1999). It seems an effective method of proselytization to expose potential members first to all kinds of social activity and only later to introduce teachings (Stark and Bainbridge 1980; Petranek 1988). This can be explained either as resulting from self-perception (people form their attitudes by observing themselves) or from avoiding cognitive dissonance, which is a term for psychological discomfort that arises when our attitudes and behavior are at odds.

tems has been selected for because of the net benefits they bring to the individual, while experessing considerable skepticism towards the group selectionist hypotheses: "If the architecture of the human mind were shaped by group selection, would religious doctrines really be necessary to promote sacrifice for the group? It appears that religious doctrines are necessary precisely because we are not likely to act for the benefit of the group when it is not in our own individual interests" (2003a, 140 – 141; also Sosis and Swartwout 2008, 299). He agrees with D. S. Wilson (2002, 93, 240) that idealized religion as presented by catechisms and such is likely to look like "pure group selection," but that one cannot equate this with the actual behavior of religious adherents. For group beneficient behavior to happen, argues Sosis, individual cost-benefit equations have to be altered in favor of it, for example by punishment. In my mind, it is in accordance with the framework proposed by cultural group selection theorists (Sober and Wilson 1998; Wilson 2002; Richerson and Boyd 2005) that we should observe a push and pull between individuals trying to free ride on at least some group benefits, while endorsing a group-level ideology that condemns such behavior in others. Selection can happen at multiple levels. For group selection to function, it requires a mechanism that favors group solidarity and punishes free riding, so that individual selection effects are dampened. Such mechanisms have been formulated by Richerson, Boyd, and Henrich (2003, 364, 368 – 369) and were presented in chapter 2.2.2.

But this leaves open the role of costly sacrifices, which empirical studies indicate to predict group longevity. Henrich (2009, 256) contends that existing mathematical models do not support the individual-centered signaling hypothesis as solving the *n*-person prisoner's dilemma,[7] but makes an interesting opening by highlighting an overlooked adaptive challenge (although one appreciated already by Rappaport 1979b, 1999): with the rise of language, alongside improved transmission of mental representations, cultural learners faced a greatly increased risk of manipulation by misinformation. Learning by mimicry, one can be fairly assured of the results, but verbal instructions are less assuring when it

---

7 Contrary to what some have claimed, the religious club model by Iannaccone (1992) does not represent a costly signaling phenomenon in the sense described above. What it shows instead is that increasing the costs of secular activities by punishment motivates individuals to contribute to religious activity instead, thus increasing its collective utility. The catch is, the rational choice theorist Iannaccone (see ch. 1.2.5) holds these religious activities to be the ends in themselves for the participants. His peacocks receive their utility from being surrounded by other long-tailed peacocks, regardless of fitness benefits. However, signals do figure as a screening device in his model regarding why individuals with different secular opportunities choose to join groups with divergent demands.

comes to edibility of foodstuffs, for one. A solution would be a "cultural immune system" that gauges how committed the model is to his or her purported beliefs. Cultural learners would be better off not trusting individuals whose actions are inconsistent with their symbolically expressed message. This sets the stage for what Henrich calls "credibility-enhancing displays (CREDs)," actions that are consistent with beliefs professed by the model and unlikely to be performed were the model to actually believe differently. Their apparent costliness can (but does not have to) depend on the accompanying message. The fitness cost of eating certain mushrooms does not much depend on whether one holds them to be poisonous or not, but sacrifices made to placate spiritual beings appear wasteful or prudent depending on one's belief in their existence.

Psychological evidence reviewed by Henrich (2009) supports this view, as people readily take communicators' self-interest into account when deciding whether to believe them. His model also shows stable equilibria with costly displays of the norm can exist, assuming the displays authenticate belief in something the learners deem highly desirable compared to the costs.[8] This in itself means little, except that the group is "locked" in a situation where practicing fitness-reducing actions has become the norm. Again, the resulting group-level dynamic depends on whether the costly displays and beliefs they authenticate are group-beneficient (altruistic, solidifying) or not and reiterates the position of Bowles and Gintis (2003, 437) that signaling and group selection may be mutually reinforcing phenomena. Group selection proliferates cooperative signaling, while signaling creates stable equilibria situations within groups.[9]

---

8 Here perceptions can matter, since we are dealing with appraisals of prestige and credibility, and track the spread of cultural representations in a population. No connection between perceptions and underlying quality is assumed, as the possible benefit is created by a representation reaching fixation and marking the group as a subpopulation committed to specific norms.

9 The need for *costly* credibility enhancing displays may also vary according to situation. Lachmann, Számadó, and Bergstrom (2001) demonstrate that even when interests conflict, communication does not have to be costly in order to be reliable. An alternative is that an honest signal is noncostly, but misdirecting signals carry a sharply increasing cost. A simple example is human language: it is "cheap talk" to speak the truth, but telling lies results in punishment. Lachmann predicts that costly signals will be favored in situations where the receiver cannot verify them within strategic time, effective social control mechanisms are unavailable, or if communicated properties are unverifiable. This has been recognized also by Sosis (2005), who argues that what is observed in closed societies should not be called "trust," as social sanctions are enough to make defection unwise. Here the more or less costly signals may work merely as highly visible markers of membership (or norm-endorsed carryovers from earlier times; see ch. 4.5). But when individuals maintain multiple group identities and the possibility of intergroup mobility is high, monitoring costs rise and sanctioning becomes less effective, favoring costly,

To sum up, the problem with Sosis' formulation of the signaling hypothesis is that it lacks a comprehensive account of how cooperativity and propensity for ritual signaling are linked within an individual. A plausible proximate explanation for the empirically observed connection is that costly ritual activity indicates internalization of relevant norms (as has been argued by Sosis 2003b) and is a public pledge of adherence to them. But not all social norms are prosocial (see e. g., Herrmann, Thöni, and Gächter 2008), so selection of cultural groups is required to sort out which of them have stabilized the most cooperative (or otherwise beneficient) norms. Thus, the connection between religion and cooperation that surfaces in empirical studies is mediated by the mechanism of religious beliefs and behaviors enforcing norm adherence and group selection pruning out both religions that enforce antisocial behavior and non-religious ideologies that have a harder time enforcing any kind of behavior. This has left the historical record peppered mostly with religious, cooperative groups, with other cases appearing as anomalous.

3.2.2 The Problem of Ideological Content

As described in chapter 3.1.4, Sosis and Bressler (2003, 227) found that the number of costly requirements predicted longevity only for religious communes, not secular ones, which led them to conclude that their theory fails "to capture some critical elements of religious belief that distinguish it from belief in a secular ideology." Other studies reviewed above point to a similiar conclusion – some element in what we conventionally label "religion" is crucial for cohesion. Three possible explanations have been offered to account for this effect: sacred moral conventions may appear more natural; religious doctrine is unfalsifiable; and the threat of supernatural sanction deters free-riding.

Sacred, naturalized moral conventions are harder to ignore than those which are explicitly man-made

Of the three explanations mentioned, this has been discussed significantly less often than the others, but was already suggested by Rappaport (1999, 164–168): "It is entertaining to note that convention becomes part of nature by being ac-

trust-promoting displays. This hypothesizing receives support from a simulation experiment by Macy and Skvoretz (1998), which demonstrates that after stable cooperation has emerged in a small community, individuals are also willing to trust strangers who carry similiar cultural markers.

counted for supernaturally." Creation myths do not only bring humans, animals, and the rest into being but also specify proper relations among them. Rituals are needed for major alterations in social information, and the conventions of other groups are often viewed as unnatural and immoral. It may also be that religious experience of "the numinous" provides a powerful source of emotional verification that contributes to an unshaking belief (Rappaport 1999, 281). Modern, secular ideologies appear more likely to stress the mutability of convention or anthropocentric abolishment of the old order and assert the right of man (and woman) to control his (and her) own destiny.[10] This could mean that individuals living under such regimes find social contracts less binding, since human-made covenants can more plausibly be altered by other humans than those set by superhuman agents. The hypothesis receives some support from the established finding that we begin to distinguish between moral transgressions (that are wrong regardless of whether they are ruled as such by an authority, law, or consensus) and transgressions of social conventions from the age of three (Turiel 1998, 903–906). As argued by Bicchieri (2006, 78), readings of situations that present a certain outcome as unavoidable make sacrifices for others easier to accept (e.g., explaining the need for cuts by blaming market forces rather than managerial ineptitude), thus constricting the leeway individuals have with a social norm.

Rappaport's argument builds centrally on the informatiotheoretical qualities of ritual acts, as described in chapter 1.2.3. One is their formality, as there is ample (independent) evidence that multicomponency (multimodality, repetition, accessory cues) of signals improves their detectability, discriminability, and memorability (Rowe 1999). Another is that regardless of whether the implied beliefs are truly internalized, public performance can be taken by others as acknowledgement of the prescribed injunctions. As the discussion of norms above (chapter 2.3) revealed, uncertainty as to what obligations have been undertaken can demolish cooperation, as their power rests on predicting (and sanctioning) others' behavior. Boyer's (2001, 248) discussion of how marriage realigns a plethora of social information (available partners, familial alliances) is illustrative of both this and of what Rappaport (1999, 52–58, 85–95) means when he says rituals can transform fuzzy, continuous information into discrete categories; while certainty of a stable

10 The French "Declaration of the Rights of Man and of the Citizen" (1789) asserts the authority of the nation over an individual, but emphasizes over and over again the freedom and natural rights of every man (here, not woman), and the same applies to "The Universal Declaration of Human Rights" (1948). While these documents naturalize these rights to an extent, they focus mostly on the positive rights of individuals, in marked contrast to the biblical Ten Commandments.

couple might otherwise take a long time to form (leaving strategic questions hanging in the air), a marriage ritual decodes this into a clear-cut distinction between "non-couple" and "couple." The individual's status becomes expressed indexically within the ritual structure, possibly aided by some more tangible displays such as gifts or sacrifices, in relation to the symbolical "canonical" content of the ritual, which gives meaning to these displays and works to situate the participants in the surrounding structural system.

Religious doctrine is unfalsifiable

A recurring suggestion by Popperian scientists (Henrich 2009, 254) is that the critical difference between religious and secular ideology is that the former contains tenets that are unfalsifiable by experience or observation. I present two criticisms of this hypothesis. First, the falsifiability of many secular ideologies is also questionable – such a complaint has been voiced against Marxism (Popper 1963, 45–46), but as the evidence above shows, socialist communes do not fare particularly well. Sosis and Ruffle (2004, 111–112) argue that the Marxist proposition "from each according to his ability, to each according to his need" must be subjectable to critial evaluation, as "the validity of this statement, namely whether this system of resource distribution is successful, can be evaluated by living according to it directives." I find this argument unconvincing, as it seems to equate the failure of a social system with the falsification of a behavioral norm. As every religiously motivated social system can also fail, and many have, as Kanter (1972, 136–137) asks, does this mean their doctrines were also *ipso facto* falsifiable?[11] Second, in day-to-day life, people do not seem to think that "confirming evidence should not count *except when it is the result of a genuine test of the theory*" (Popper 1963, 48, emphasis his). To the contrary, there is ample evidence of a psychological confirmation bias, our tendency to unconsciously seek verification for one's beliefs and disregard negative or refuting evidence against them (Nickerson 1998). If so, it is difficult to see how the (socio)psychological efficacy of a belief could be substantially improved by its unfalsifiablity.

It seems to me that comprehensive ideologies (religious and political) resemble Kuhn's paradigms more closely than hypotheses or theories, and are not sub-

---

11 A more subtle distinction can be made between deontic and ontological doctrines, but it is by no means clear which ones dominate in any given tradition. It has often been pointed out that religious believers are less concerned with cosmology or abstract dogma than outsiders believe them to be, and more concerned with practical questions of managing everyday life (Nickerson 1998, 209; Boyer 2001, 137–138; Wilson 2002, ch. 3).

ject to direct testing, even if some parts of them are falsifiable. But Sosis (personal communication) suggests that the crucial difference is in the way falsifiable norms are embedded in the ideology. Rappaport's (1999, 263–268, chapter 9) hierarchy of liturgical orders ranges from "Ultimate Sacred Postulates" (such as creeds) that lack material significata and thus falsifiability, through "cosmological axioms" that ground social distinctions into transcendent concepts, to social rules (norms) that explicitly dictate behavior in daily life, and because of this materiality, are subject to testing.[12] Thus, while these rules are subject to failure, and the axioms can change with prevailing historical or environmental conditions, the sacred postulates remain and can also sanctify new, better-adapted axioms and rules.

To illuminate what this might mean, consider Henrich's (2009, 254–255) example that "various groups have come to believe that faith, or a ritual, can provide protection from bullets. Such beliefs have tended not to endure for long periods, once the shooting starts." Incidentally, when the British invaded Tibet in 1903–1904, some Tibetan troops had been presented with the Dalai Lama's personal seal, which their priests promised would make them bullet-proof. The ensuing massacre refuted these promises; a journalist's report tells how instead of running, retreating Tibetans "walked with bowed heads, as if they had been disillusioned in their gods." However, subsequent events (and even fiercer resistance) do not indicate general defections from Lamaism (Hopkirk [1982] 1995, 173–180). It may thus be, as Sosis and Bressler (2003, 228) believe, that the critical difference between religious and secular rituals is that the latter lack references to supernatural entities, which makes them less likely to weather challenges. This connects with the suggested explanation above, that religious rituals are grounded in more naturalized conventions. But spirits and deities may also have another role in upholding morality, as will be shown in the next section.

Belief in omniscient, moralizing agents makes trespassing less attractive

This suggestion, made first by Voltaire (*"Si Dieu n'existait pas, il faudrait l'inventer"*) and more recently by several scholars (Johnson and Krüger 2004; Johnson 2005; Shariff and Norenzayan 2007; Johnson and Bering 2009; Shariff, Norenzay-

12 It should be noted that Rappaport (1999, 265, 287–289) is very explicit in stating that he does not claim that such hierarchies constitute logically derived systems. Cosmological axioms are axiomatic, not derived from the USPs, which rather work to sanctify the whole construct while remaining beyond tests of logic and experience. This is important, since attempts to formalize theologies into logical systems have repeatedly failed or produced results far removed from how most of the religious understand them.

an, and Henrich 2009), has the appealing quality of applying to what is in many studies taken as the hallmark of religious groups: gods and spirits. As reviewed in chapter 3.1.2, cues of being monitored reduce defections (Bering 2004, 425–426; Bering, McLeod, and Shackelford 2005; Haley and Fessler 2005; Bateson, Nettle, and Gilbert 2006; Burnham and Hare 2007; Shariff and Norenzayan 2007), and those societies that believe in moralizing deities are larger and possibly more cooperative (Snarey 1996; Roes and Raymond 2003; Johnson 2005). In view of this evidence, religions appear as complex systems (chapter 1.2.1) that are more than the sums of their parts. The emergent properties that manifest as longevity and cohesiveness disappear if one crucial part – costly rituals *or* belief in the supernatural – is removed. But Schloss and Murray (2011, 48–51) have pointed out that there are actually two distinctly separate versions of the supernatural punishment hypothesis that both claim to account for the empirical data. The "cooperation enhancement" account predicts that the idea of constant monitoring would reduce defections, which has led to the evolution of such belief-systems on the cultural group level. The "punishment avoidance" account posits that if the costs of being punished for defecting are considerable, beliefs that reduce cheating are adaptive for the individual.

The cooperation enhancement hypothesis (e.g. Shariff, Norenzayan, and Henrich 2009, 130–132) builds on the observation that over historical time, religious systems have evolved towards featuring omniscient deities with interest in human morality and behavior ("high gods") and punishments meted out in the afterlife (whether this constitutes reincarnation or transcendent realities). Such shaping of the social selective environment may have also enticed a genetic response, making commitment to such gods easier, but the logic of the argument does not depend on this. While straightforward and plausible, the problems of the argument lie in the hitherto unexplored gaps between belief and (costly and non-costly) ritual action and the overall validity of cultural group selection accounts (Schloss and Murray 2011, 53–54). The present work offers some suggestions on reconciling these.

The punishment avoidance hypothesis can be formalized as equation $pc > m$ (Johnson and Bering 2009): if the costs of punishment (c) weighed by the probability of detection (p) exceed the costs of missed opportunities (m), avoiding mistakes by overestimating p is a winning strategy and leads to the genetic evolution of "god-fearers" over "Machiavellians" (who defect when they think they can avoid punishment). Haselton and Nettle (2006) illustrate with both models and empirical cases how it is often a better adaptive strategy to maximize self-protection over resource intake by overreacting to potential threats. This idea is similiar to the hyperactive agent detection device hypothesis advanced by cognitive scientists of religion (Barrett 2004; also Guthrie 1993; Boyer 2001), but it posits an adaptive function to this bias. Schloss and Murray (2011, 55–57), however, identify several problems with the inbuilt assumptions of this logic. It builds centrally on the assumption that human overconfidence in our abilities needs to be selectively damped. Belief in high gods may or may not be the (evolutionarily) best solution to this problem. But as it appears to arise in conditions where large society size makes managing cooperation by reciprocity, reputation, and punishment unwieldy, this argument runs counter to the original hypothesis where this belief is an adaptation precisely to conditions of efficient punishment. This appears to be a serious problem for the present formulation of the hypothesis.

The commensurability of the two hypotheses is debatable (see Johnson's 2011 comment, 82–83), but neither is very clear on the relationship between deities and rituals. Boyer (2001, 150–160) characterized gods and spirits as "full access strategic agents," that is, as possessing more or less of the relevant knowledge (or "gossip") about the behavior and dispositions of individuals. As humans are not capable of representing endless chains of information about how others perceive things (such as changes in social relationships), Pyysiäinen (2009, 38–40) suggests full-access strategic agents are an effective way of representing collective knowledge. He predicts that such agents are necessarily a feature only in rituals that bring about changes in the "shared" knowledge of a group, such as transformations of social positions.[13] This may be because the thought of a norm or principle not being laid down by someone is unnatural to us, a suggestion that ties in well with the above account of supernaturally naturalized conventions.

Religious traditions are multifaceted, and learning to live by them requires some dedication. Peters (1980, 3–4) claims that "[t]hroughout the Middle Ages and early modern European history, theological uniformity was synonymous with social cohesion in societies that regarded themselves as bound together at their most fundamental levels by a religion." Opposition in beliefs meant opposition to a whole culture regardless of specifics, whereas Jews and Muslims did

---

13 This prediction follows the ritual hypothesis laid out by Lawson and McCauley (1990; McCauley and Lawson 2002), who predict that there is a distinct class of rituals (initiations, weddings, funerals) where the "culturally postulated superhuman agent" is seen as directly connected to the agent performing the ritual, which gives it legitimacy. Malley and Barrett (2003) interviewed Hindu, Jewish, and Muslim students to test whether they had the ritual "competence" posited by the Lawson-McCauley theory. Subjects named rituals from their religion and were asked questions concerning their properties. While the predictions of the theory were carried out clearly above chance, the informants were often unsure about the role superhuman agents had in a ritual.

not constitute dangerous heretics in the same sense. Bulbulia (2004) and Mahoney (2008) have suggested that theological expression in itself would constitute a costly signal in the form of "ostentatious religious expression" and "self imposed Zahavian handicap" that signals intellectual and economical capacity (cf. Veblen 2004; Spence 1973). But in viewing the suggestions of these authors, it is hard to avoid the impression that their conception of theology is somewhat biased. The "counting angels on pins" example given by Bulbulia and often attributed to scholastics is apparently a contemporary "hostile jibe" not found in any of their own writings (Lang 1992, 284), and Mahoney quotes from the Gnostic Christian *The Sophia of Jesus Christ*, an exceedingly mystical piece, to illustrate that "[t]heological information is often presented in almost incomprehensible language" – hardly representative examples. A rather divergent analysis is offered by D. S. Wilson (2002, chapter 3), who finds in John Calvin's catechism a virtual guidebook for the creation of an adaptive community.

So, instead of trying to find one or two structural features that would allow as to account for the success of religious communities, we may have to consider the whole package – how a combination of purposeful design and historical selection pressures has created combinations of norms, beliefs, and practices that integrate individuals into a whole. To pick a novel example, many religions (and armies) incorporate singing, dancing, and marching into their institutions, synchronous activities which Wiltermuth and Heath (2009) found to enhance cooperativity between participants regardless of emotional involvement. Kanter (1972, 121) notes, "[T]he successful communities tended to provide programs and philosophic guides for much of the behavior of members, whereas the unsuccessful ones did not." The number of ritual requirements, for one, appears to provide a good heuristic with which to predict group longevity, but any badge or ban may not be as good as the next one in creating a cohesive and functional group. Richerson and Boyd (2005, 180–186) offer an illuminating analysis of the Amish. Although known for their Luddite traditionalism, they take advantage of modern medicine and live integrated with the surrounding economic system. There is a reason they shun cars and telephones: these technological advances would allow living far away from fellow Amish, weakening the group as a unit. And instead of counting their peculiar executive system as just another costly requirement, it is more insightful to note that it appears designed to prevent internal competition for positions of power.[14] These are the kinds of safeguards that

14 It should be noted that not every costly obligation is necessarily a costly signal in the sense that it measures an individual's commitment to group norms. Some obligations work to constrain harmful individual behavior (as in this case) or direct the evolution of the group. To pick another example, during the sixteenth century Protestant Reformation infant baptism became a

can uphold the between-groups selection pressure against the inside-groups pressure.

3.2.3 The Problem of Emotions

In chapter 2.4 I described how emotional expression could function as a reliable signal of commitment to particular beliefs and behaviors. This has led some authors (Murray and Moore 2009, 240; Bulbulia 2012) to question why costly (religious) signals are required if emotional (religious) signals achieve the same effect for free. Although the cheapness of emotional expression is questionable (as will become evident in my case study of Finnish revivals below, socially deviant emotional expression is anything but free), I focus on three critiques that highlight the differences between emotional and ritual signaling.

First, Irons (2001, 293) already remarked that while emotions are fine and necessary, culturally specific substance is needed to signal commitment to a particular group or society. While plausibly honest, an emotional display can be motivated in seconds regardless of past loyalties. Continued ritual participation and familiarity with local ways indicates more sustained loyalty (with permanent markers such scars and tattoos as the limiting case; see Sosis, Kress, and Boster 2007). As mentioned above, during the American Second Great Awakening some preachers put their enthusiastic converts on probation for four to six months before being accepted into their church (McLoughlin 1978, 122).

Second, it is useful to consider what kinds of signals are easiest to monitor. Iannaccone (1994, 1188) noted that while it is difficult to constantly observe whether an individual participates in a required activity, it is easier to spot even single instances of forbidden acts, such as the breaking of celibacy, food taboos, or dress code. And while injunctions on specific emotional displays (e.g., disgust at ritual paraphernalia) are certainly possible, emotions make bad badges or markers, as they are not worn constantly. Also, Bulbulia (2012)

topical issue, with some radical "anabaptist" groups (from whom the Amish and Hutterites happen to descend) rejecting it in favor of baptizing only adult converts. While taking this second baptism can certainly be interpreted as a costly signal of commitment (at least in the prevailing conditions of persecution), with infant baptism it is less so. But even more important is to note how these two differing norms point the corresponding groups to very divergent evolutionary paths: infant baptism in a state religion where every member of the society at least nominally professes the same faith (as had been in western Christendom since the time of Theodosius I); adult baptism in an exclusive community of believers which exists in opposition to the surrounding populace. It was for this reason that Martin Luther, among others, fought for infant baptism (MacCulloch 2003, 143).

considers the possiblity that rather than functioning as signals of commitment themselves, "rituals are established to elicit emotional evidence of anthropomorphic commitment," with their costliness a by-product. True, some level of emotional valencing is involved in rituals across cultures (Alcorta and Sosis 2005, 332–340), and costliness would seem an effective way of bringing emotions to the fore, especially if pain or involuntary sacrifice is required. But research on Whitehouse's "modes of religiosity" theory illustrates that there is substantial variation in the degree of this emotionalism, which is also connected to other sociocultural variables rather than just to ritual acts (2000; Whitehouse and Martin 2004).

It would seem better to seek a dynamic relationship between emotional and costly signaling. Schloss (2008, 201–203) points out that the trouble with costly signals is that if they do work as investments with calculable payoffs, shrewd individuals can undertake them without committing to the group or its ideology. Committed individuals stand to suffer when such "hypocrites" decide it is more opportune to defect. He hypothesizes emotions to solve this problem, as they are harder to display if one is only in it for the money. Emmons and McNamara (2006, 15, 23) suggest the feeling of gratitude to be the proximate mechanism that regulates reciprocity, and that in religious contexts it can "eschew finely calibrated tit-for-tat calculations of cost-benefit," as reciprocation of incalculable divine gifts is impossible, leading to extreme altruistic self-sacrifice as evidenced by the lives of saints and martyrs. As a closely related issue, Atran (2010, 341–344, 379–380, 395) reports that attempting to resolve territorial conflicts over "holy lands" (Palestine, Kashmir, the Black Hills) by offering material incentives tends to backfire, as the involved parties feel repulsion (that scales up with the offer) over the suggestion that their sacred values could be bought. The cost-benefit logic applied by subjects in these situations does not follow a rational (maximizing) calculus.

This hypothesis leads to the prediction that suspicions of hypocrisy and insincere belief (professed for material advantages) should make emotional displays (and demands for them) more prominent in religious life. As Schloss notes, at least in the Western world dramatic renewals of religious commitment have been characterized by manifestations of phenomena such as glossolalia, fainting and trembling, fits of trance and ecstacy, and passionate weeping and laughing (for examples from American revivals, see McLoughlin 1978, 61, 84–87, 123–124, 136). They convey in a manner that is hard to deny that the subject has experienced another reality. Schloss takes Pentecostalism as a salient example of such a movement, and a comparative study does indicate that Pentecostals report trust in their coreligionists significantly more than people in other Christian communities included in the study (Lugo et al. 2006, 49–51).

As will be seen below (chapter 6.2.2), in early nineteenth century Finland the revived believers accused their parish ministers of preferring socializing with other gentry over their religious commitments, which they performed in too mechanistic a fashion. In contrast, the preachers from their own ranks showed open emotion with their pastoral duties. The revived would shun those members of the clergy who had not experienced such a spiritual awakening and so formed their own separatistic congregations (Heino 1976, 158–162; Raittila 1976, 206, 235; Ylikangas 1979, 120–123, 135–136, 153–156; Lohi 1997, 658, 665; Toivanen 1997, 78–89; Hirvonen 2003, 37–38; Peltonen 2006, 231; Kimanen 2008, 83–84, 104, 165–169). We can interpret such behaviors as a strategic decision to limit cooperation to those whose commitments can be trusted to be genuine. Once individuals become suspicious of being duped, the first reaction should be to limit interactions with the guilty parties, effectively dismantling cooperation. Formation of a new, more tightly-knit group can serve both to compensate for the missed opportunities and to offer security against possible retaliation.

However, such emotionalism appears short-lived; even though many religious movements start as intensely emotional affairs, with time the emotional element becomes more and more muted. The process can be compared with the well-known psychological phenomenon of emotional evanescence, the tendency of emotional reactions to both positive and negative events to wear off quickly. Such "cooling" is not surprising, as intense emotions disrupt our ability to function, and regaining functionality quickly is important. A hypothesized mechanism for this is the process of making sense of novel events to make them appear more predictable (Wilson, Gilbert, and Centerbar 2002). This turns extraordinary events into ordinary ones and robs them of their emotional power. Thus, in the context of revivals it can also be expected that as individuals begin to routinely apply their new beliefs in understanding the world around them, they are unable to react with comparable emotion. Concurrent emergence of new norm systems may actually work to make this world more predictable, as social norms have been found to resemble self-fulfilling prophecies of how others act in response to our behavior (Ylikoski 2011, 52; chapter 2.3). However, this may be enough: even a short-lived emotional outburst could plausibly work to help individuals reliably identify those who genuinely hold to their professed beliefs and help form a new group based on this. After this, other kinds of institutionalized practices that help demarcate those who supposedly share the same norms can work to direct prosocial punishment of defectors and maintain group cohesion. I will further discuss this matter and other generational effects below (chapter 4.5).

3.3 Chapter Conclusions

In this chapter I have attempted to demonstrate that there exists between religiosity and prosocial behavior a robust, predictable relationship that is mediated by adherence to social norms. On an individual level it manifests as greater trust in perceived coreligionists and actual acts of altruism, most saliently when their benefits are confined to ingroup members. On a group level this has the consequence that religious groups are more stable (that is, long-lived and cohesive) than their less religious counterparts.

This religiosity is best measured by belief in moralizing agents with wide-ranging observational powers (deities and spirits) and behavioral standards that demand individuals contribute time or resources (most saliently ritual requirements). The theoretically most plausible interpretation is that these work to empower norm observance, as engaging in sacrificial behaviors authenticates to observers a person's belief in their essential nature, and deities with an interest in human morals can cause individuals to refrain from norm-breaking for fear of divine sanctions.

Fears of hypocrisy can be countered by emotional expressions that validate sincere experiences of the otherworldly. A community demarcated by these practices is in a privileged position to recognize fellow members who share the same norms and can work together within their dictates, sanctioning those who lapse. It is, however, the complex whole of social norms and institutional arrangements that determine eventual group success or failure (relative to others), not any single quantifiable factor.

4 Deprivation, Hazards, and Religious Revivals

> The fourth generation, then, is inferior to the preceding ones in every respect. Its member has lost the qualities that preserved the edifice of its glory. He despises those qualities. He imagines that the edifice was not built through application and effort. He thinks that it was something due his people from the very beginning by virtue of the mere fact of their descent, and not something that resulted from group effort and individual qualities. (Ibn Khaldûn 1967, 106)

> All this has its origin in group feeling, which differs in different groups. Luxury wears out royal authority and overthrows it. [...] Eventually, a great change takes place in the world, such as the transformation of a religion, or the disappearance of a civilization, or something else willed by the power of God. (Ibn Khaldûn 1967, 115)

As seen in previous chapters, theoretical studies suggest that ritual behavior and belief in deities are potentially adaptive, as they work to enhance norm adherence, which in many cases reduces the costs inflicted by defectors and those from other monitoring arrangements. Empirical studies give strong evidence that religious groups are indeed more cohesive than comparable secular groups. This leads to the question, what prompts the formation of a religiously motivated group, and how does it happen? For one, if religiosity (as a complex of coalesced, cognitive, and behavioral tendencies such as committed belief in the supernatural, ritualized behavior, and altered states of consciousness) is an adaptation to selective pressures (Sosis 2009, 343–344), we might be equipped also with a mental module that responds to a reading of environmental factors with marked religious behavior, with the aim of signaling social need to prospective partners. The writings of Alcorta and Sosis (2005, 2006) and Bulbulia (2004), with their emphasis on the fitness benefits of religious behavior for an individual, suggest this scenario. And many aspects of religious behavior do seem to converge on enhancing cooperativity, as seen above; even praying for another person has been found to increase forgiveness when compared to control conditions (Lambert et al. 2010).

Alternatively, group formation can happen in a more haphazard manner, initiated by less adapted responses to hazards. For example, Gray and Wegner (2010) hypothesize that belief in supernatural agents stems from a need to find a moral agent to account for suffering. Several studies have also documented magico-religious responses to increased death salience (Keinan 1994; Norenzayan and Hansen 2006; Sosis 2007), behavioral tendencies which may be byproducts of an evolved hazard-precaution system (Boyer and Liénard 2006; Liénard and Boyer 2006; see below). But when made public, these representations spread quickly in a social network because of their cognitive attractiveness (Sperber 1996; Boyer 2001; Henrich 2009). Individuals also adjust their representations to match those of others. A recent study by Stroope (2011) found that an American Christian's affirmation of biblical literalism is connected not only to his or her own level of education, but also to that predominating in his or her congregation; surrounded by less-educated people, even a highly educated person is more likely to be influenced by more literalist views (and vice versa). If these processes create interlocking patterns of sociocultural behavior (i.e., the individuals who exhibit matching public representations feel they share a common identity, norms, and concerns), the stage is set for cultural group selection, where higher internal cohesion and norm-observance gives religious groups a competitive edge. This advantage would result from the features described above: clearer demarcation of the in-group (enhanced by communicative properties of rituals), emotional manipulation, cooperativity-enhancing features of synchronic singing and movement (Wiltermuth and Heath 2009), and belief in constant monitoring by invisible agents.

While this view confers potential adaptive benefit to religious behavior, it is less clear whether religion constitutes an actual adaptation (see Sanderson 2008, 145–147 and Sosis 2009, 345 for clarification of the conceptual difference). I am sketching these alternative scenarios to illustrate that treating religious displays as a potentially adaptive response to survival hazards does not commit us to either side on the current adaptationist-byproduct debate (Boyer and Bergstrom 2008; Sosis 2009). A potential resolution to it is the exaptation view offered by some authors (Pyysiäinen 2010b, 12; Pyysiäinen and Hauser 2010, 105; Schloss and Murray 2011), in which it is considered a distinct possibility that the cognitive capabilities involved in religious behavior first arose as adaptations to selective pressures that had little to do with religion or cooperation. Once in place, these pressures were co-opted for building religious systems that resulted in increased cohesion. Such a view also fits unproblematically with the cultural group selection framework, which does not preclude a genetic response when those best endowed to cope in the new sociocultural situation have thrived (Henrich 2004, 30; Richerson and Boyd 2005, chapter 6). This would have reinforced some or all of the abovementioned tendencies, or made the parts act in a more integrated manner (cf. Sosis 2009, 344). A salient example of this would be the fact that religious beliefs, despite their appearance of being low-cost misfires of more basic cognitive systems, are not muffled but instead amplified with strong emotional valences, which suggests that selection has been acting on them (Bulbulia 2004, 21).

Certainly the process of *group* formation also includes a strong cultural level component, as current accounts on the birth of new religions emphasize the slow, negotiatory character of their inception and maturation. It is not only the beliefs and institutions of the new group that are in their formative stages, but also the precise location of the boundary between the in-group and the out-group. For example, while Christianity was born in a Jewish context, it took a long period for adherents to form strictly separate identities as orthodox Christians and rabbinic Jews. According to Boyarin (1999, 5–6), up to and during Late Antiquity (fourth century), various forms of these religions shared many ideas and contended with one another for primacy. The negotiatory process is captured in the church fathers' complaints regarding believers in Christ who worship in synagogues and follow Jewish ways (Boyarin 1999, 11, 17). Donner (2010, 69–71, 83, 110–114, 203, 217, 221–223) offers a similiar account of the emergence of Islam: that the original community of believers around Muhammad was a monotheist reform movement that included also "righteous" Jews and (non-trinitarian) Christians and whose militancy was directed against impiety and unbelief in God's oneness, not against these other faiths per se. Islamic polemic against them appears only by the late seventh century, when the community had outgrown its early expansionist vigor and when a stable empire identity built around waging war against the sinful world of unbelievers was no longer sustainable.[1]

A complete discussion on how ideological borders between religions are negotiated is beyond the scope of the present study, and I now turn my attention back to the micro-level causes that prompt group formation. To frame this discussion I introduce two theoretical constructs that have been employed to explain behavioral time changes in religiosity: the deprivation theory and the hazard-precaution system model. Subsequently, I proceed to review empirical literature on religiosity as a survival strategy in changing environments. After this I am able to make some suggestions about what kind of dynamic can be observed and how the signaling hypothesis can be employed in explaining religious change.

4.1 The Deprivation Theory

Marx ([1843] 1977, 131) called religion "the sigh of the oppressed creature" that is both an expression of misery and a protest against it. He thought it provided illusory happiness for people in search of true happiness. Later, in the sociology of

1 Still, the perception of Islam as a Christian reform movement (that appears for example in the writings of St. John of Damascus) took some time to dissipate, as Jenkins (2008, 191–192) has noted that even the eighth century Byzantine iconoclasm appears to have been prompted by the Muslim caliphs' strict prohibition of religious images in the 720s, indicating that the rigor of the "reformists" still had the power to shape orthodox Christian practice.

religion, H. Richard Niebuhr ([1929] 1987, 18–20) took the church-sect distinction of Weber and Troeltsch and proposed that unfavorable social conditions result in the formation of sects (separatist religious groups that reject the values of the surrounding society) that are gradually transformed into churches (inclusive religious groups that encourage conformity). Speaking of Protestant history, he argued that sects tend to draw their support from the poor and the outcast. However, with the coming of the next generation, the sect is gradually transformed into a church, as the children rarely share the fervor of their parents and as "the discipline of ascetism" increases members' wealth.

Building on this, Glock (1964, 27–29, 34; similiarly in Glock and Stark 1965, chapter 13) identified five different kinds of deprivation that can cause the formation of movements: economic (income differences, causing primarily sect formation); social (want of status and opportunities for participation); organismic (mental and psychical health); ethical (losing faith in the values of society); and psychic (problems with the meaningfulness of life). Glock conceived deprivation to mean "any and all of the ways that an individual or group may be, or feel, disadvantaged in comparison either to other individuals or groups or to an internalized set of standards." While suggesting that a shared feeling of deprivation (mainly of the three former types) is a necessary precondition for the rise of any social movement, he maintained that it results in the formation of religious groups mainly when "the nature of the deprivation is inaccurately perceived or where those experiencing the deprivation are not in a position to work directly at eliminating the causes," as religion is likely only to compensate for feelings of deprivation instead of efficaciously working at their causes (this idea of religious compensators was further elaborated by Stark and Bainbridge 1996, 36–42). But it was not only sect formation but also the survival of churches that Glock attributed to feelings of deprivation, as he saw them as dependent on the condition of social deprivation, without which their members would have no need to seek compensation.

By 1979, Hoge and Roozen noted that "[t]he deprivation theory is the most common theory of religiosity found in the literature," being called "the most 'standard' theory of religion in the history of social science" (Hoge and Roozen 1979, 48–53). However, most of the empirical studies they reviewed had found either no relationship or a positive relationship between socioconomic status and church attendance (see also Roof and Hoge 1980). So the stability of churches can clearly not depend on continued deprivation. As the church-sect theory predicts socioeconomic upward mobility for sect members (Stark and Bainbridge 1985, chapter 7; e.g., Isichei 1964), it could still be argued that sect formation and joining happens in conditions of deprivation (Aberle 1965; Lofland and Stark 1965; Simmonds 1977; Harris 1978, 97–127). This leads to an oft-mentioned theoretical criticism: both the deprivation theory and the church-sect theory fail to supply a viable mechanism for this transformation. To remedy this, Stark and Bainbridge (1985, 153–154) advanced a hypothesis based on statistical regression towards the mean with generational change.[2] This may well be a contributing factor, but as there seem to be no studies documenting a *downward* mobility within the high-SES churches, there would also have to be at least some constraining factor that hinders mobility in this direction.

Another criticism applies to the evolution of sects into churches or denominations by the second generation, as assumed by Niebuhr, which B. R. Wilson (1990, 107–108) considers empirically ungrounded and insensitive to contextual variation. The loosening of rigor and separatism associated with denominationalisation does not happen with every sect, and with some goes only halfway. The determinism and context-insensivity implied in the deprivation theory appears misguided. The concept of path-dependency as employed in the field of institutional economics (North 1990, chapter 11) goes some way toward explaining why an institution is prevented from evolving in a direction that would make it more adapted to its environment or produce more benefits to its constituents: it is constrained by previous developmental choices (its history and culture). For example, the Jehovah's Witnesses' conspicious lack of relaxation may be attributable not only to the development of strong control by central bureaucracy (Penton 1997) but more specifically to their practice of completely shutting out both transgressors and non-Witness contacts. Little tolerance for testing the limits removes those willing to experiment with the tradition, while the core belief of suspicion towards the dangerous outside world attracts those with an already pessimistic worldview (Holden 2002a, 4; 2002b, 69–71, 79, 144, 163). Their constituency changes rather than their institutions or beliefs. But it is harder to argue that some path chosen inevitably leads to loosening rigor without embedding this into a theoretical framework that predicts such relaxation unless constrained. Niebuhr's original insight about wealth accumulation may still be correct but has been under-appreciated for lack of proper theoretical formulation.

2 Regression towards the mean is a phenomenon that occurs if: 1) the initial sample is over-recruited from the low or high end of distribution pertaining to some trait in the population; 2) there is a tendency for the trait to cluster around the population mean; and 3) generational change is not wholly deterministic, i.e., there is some variation in replication. In this case, Stark and Bainbridge (1985) argue, even if initial sect members tend to have for example low IQs, their children will by chance only have IQs closer to the population mean.

4.2 The Hazard-Precaution System

More recently, Boyer and Liénard (2006, also Liénard and Boyer 2006) have built a case of ritualized behavior (whether cultural, developmental, or pathological) stemming from a general-purpose "hazard-precaution system" geared to detect and react to inferred fitness threats. They argue that the compulsiveness, rigidity, redundance, and obsession with purification and danger that characterizes these groups results from an evolved concern towards specific threats (see below) and an arousal state in which non-action feels dangerous. Cultural, collective rituals are seen as derivative of individual ritualized behavior and resulting from attractor-based cultural selection (following Sperber 1996). Features in cultural rituals, from rigidity and purity-obsession to implied danger of nonperformance, correspond to the aforementioned traits and make such rituals attention-grabbing. Theirs is thus a by-product account. Alcorta and Sosis (2006) argue vehemently against this interpretation of the evidence, noting that the core elements of ritual, such as formality and repetition, do actually work to focus attention and enhance learning (Rowe 1999) instead of merely swamping the working memory. This could result in strengthened associations that facilitate non-agonistic behavior with individuals involved in the signals and symbols of the ritual. Consequently, Alcorta and Sosis (2006) view this capacity as an evolved adaptation.

As sketched above, it does not necessarily matter here whether we have a brain module dedicated to generating religious responses or if it is a question of by-product behavior and cultural attraction. Both require mapping from cues of survival hazards to public behavior, with a cognitive system in between. The subjective experience of these hazards is no doubt mediated by a number of different cognitive systems that specialize in inferring the presence and severity of a particular threat. Boyer and Liénard (2006, 603) introduce an indiscrete division of threat situations into those where the danger to fitness is manifest and those in which it has to be inferred from less explicit clues. It is the latter where a system for parsing and generating responses to potential danger is required. Also, the better-safe-than-sorry nature of many threats should result in the system being quick to activate and slow to power down.

Boyer and Liénard (2006, 602–603; Liénard and Boyer, 820) review evidence of recurrent hazards of our likely evolutionary background being more salient than others, regardless of prior exposure. But most or all of these are threats that can be substantially reduced with committed collective action:

- *Contamination and contagion from pathogens:* While these threats may often be beyond pre-modern methods of social action, even basic care of food and water increases the chances of survival with many diseases (McNeill [1976] 1998, 135). Consequently, Stark (1997, chapter 4) depicts the willingness to nurse the diseased during epidemics as one of the group-beneficent traits of early Christianity.

- *Reproductive risk, such as low-fitness partners, un-nurturing males, cuckoldry, and harm to offspring:* To pick one of these, while ritual precautions imposed on menstruating women would seem to be a good candidate for resulting from hyperactive aversion to contagion, there is a study by Strassmann (1996) that shows how among the Dogon of Mali it functions as a male-imposed anti-cuckoldry strategy with the exactitude of hormonal tests (of course, this behavior is adaptive only from the perspective of individual males).
- *Resource scarcity:* This may often be a question of solving public good dilemmas in order to effectively employ sharable resources and to avoid a "tragedy of the commons" (see Sosis' studies on kibbutzim). Sundberg (1993, 164–168) concluded in her study of early modern Swedish communities that with increasing competition for available resources, social relations all became more decisive in their usage. Hayden (1987, 83–84) claims that early hunter-gatherers coped with episodic resource scarcity by establishing alliances with other bands whom they could "visit" while their own area suffered from adverse conditions. Summing up the anthropological perspective, Douglas (1986, 27) states that factors favorable to collective action have "to do with the ratio of population to resources, together with a possibility of satisfying wants without engaging anyone in the hard, monotonous, sustained kind of work that tempts some to coerce the service of others." I will comment more on this issue below.
- *Predation and intrusion by strangers:* Seeking safety in numbers can directly reduce these threats. Pyysiäinen (2008) has suggested that the hazard precaution system might be connected to territorial thinking in humans, especially as evidenced by the human tendency to treat borders as sacred and ritually enforce this (Anttonen 1996). Furthermore, a simulation experiment has shown that trust in strangers can emerge from shared cultural markers such as religion (Macy and Skvoretz 1998).
- *Social harm, such as ostracism and reduced cooperation:* As these are problems of social action, any method of restoring mutual trust and commitment to the group would seem to have potential adaptive value in confronting them.

The deprivation theory predicts a religious response to deprivations when their source is misperceived. The hazard-precaution system model is based on the assumption that our neurocognitive systems are easily activated by cues that are (mis)taken to imply danger. But it does not matter so much whether the individuals involved accurately perceive the source of their troubles, as a body of individuals organized into normative behavior can overcome all sorts of obstacles. Not every new movement is a success (Stark and Bainbridge 1985, 126–134), but those that are can substantially alter the living conditions of those involved by introducing a network of trustworthy agents, mutual social insurance, or resource-sharing systems (Isichei 1970, 185–186; Kirby 1993; Ensminger 1997; Berman 2000, 911–912; Paxson 2004). Here, the commitment theories of religion fill the missing link in the deprivation theory, showing how sect members' socioeconomic status can be transformed.

4.3 Religion as a Survival Strategy

There exist a number of studies that give systematic evidence of religious behaviors and institutions providing a survival strategy in conditions of varying threats. Analyses by Norris and Inglehart (2004) demonstrates a strong negative correlation between religiosity (participation, values, and beliefs as measured in World Values Surveys) and socioeconomical development (measured by The Human Development Index provided by UNDP) across 76 nations. From this they argue that secularization results from existential security experienced in developed countries, with vulnerability to risks and mortal danger in formative years being especially decisive.[3] Their prediction is that societies with few doctors and infrequent access to improved water sources are much more likely to generate a heightened demand for religion, as is happening in many developing countries. Also, in a brief examination, Gray and Wegner (2010, 11–12) found a strong and significant correlation between per-state indices of human suffering and religiosity across the United States.

Previously, Hayden (1987) has demonstrated with contemporary hunter-gatherer data a significant link between resource stress and belief in celestial deities that he argued would work as emotionally powerful unifying concepts in between-groups interactions. Also, Snarey (1996) used *Standard Cross-Cultural Sample* and found a strong and highly significant association between belief in a supreme deity concerned with human morality and scarcity of water, a resource critical to group survival. Based on archaelogical data, Lernia (2006) interprets

---

3 The traditionally problematic case of the United States for secularization theories is explainable by the high economic inequality and lack of welfare safety net in the country. Norris and Inglehart (2004) also found a positive correlation between religiosity and faith in the benefits of science. This would suggest that secularization does not result from scientific progess unless it is employed to create welfare systems.

the rapid spread of cattle burials in megalithic structures as a social response to abrupt climatic changes (increased aridity) in the Holocene Sahara. He further suggests that the later transition to burying humans instead of cattle may conincide with another period of dramatic environmental change for the worse.

There are also relevant studies concerning the Ghost Dance movement of 1889, which involved several (but not all) North American native tribes and has generally been viewed as stemming from deprived circumstances (Smoak 2006, 115, 118). Carroll (1975) found support for two hypotheses: 1) tribes that at the time had only a short time (less than ten years) to adjust to the extermination of the buffalo (entailing both economic deprivation and the end of culturally significant activities) in their area were more likely to have accepted the Ghost Dance than those with a longer period of adjustment; and 2) tribes with unilineal descent groups were less likely to have accepted the Ghost Dance than those without (Carroll reasons that unilineal descent is more likely to produce discrete, integrated kin groups with committed members). Both of these conditions appear to have been sufficient for the acceptance of the movement. In addition, Thornton (1981) found tribes that had experienced short-term population decrease to have been more likely to participate than those that did not. The argument that revivals are spurred by a need for committed cooperation receives support from the finding that existing institutions of commitment may have inhibited it, a reasoning that follows closely that of Carroll's (1975, 397).

Some studies have documented a religious change in response to fluctuations in economy. Analyzing conversion rates of American churches during the periods 1920–1939 and 1961–1970, Sales (1972) found that when recession strikes, conversion to authoritarian churches increases, while strong economic times favor nonauthoritarian ones. Building on this study, McCann (1999) confirmed the result with annual data from 1928–1986 and threat indices that also included social and political issues in addition to economic ones (finding that the observed effect is best explained as resulting from the aggregate threat levels).[4] Also, Padgett and Jorgenson (1982) found economic threat to predict the level of superstition (as measured by the number of articles on astrology, mysticism, and cults in a periodical index) in Germany from 1918–1940. On the other hand, increasing benefits can also drive up commitment costs. Berman (2000) shows how after the Israeli government made Yeshiva school attendance grounds for draft deferment, the Ultra-Orthodox Jews begun spending even lon-

---

4 It should be noted that the churches designated as "authoritarian" are characterized by demands on discipline, conformity, and commitment (McCann 1999, 330–331), which correspond to what is meant by costly requirements in the present study.

ger times in these schools, well past the draft exemption age. His argument is that in order to remain part of the Ultra-Orthodox community, the school attendees need to demonstrate that they are motivated by religious concerns, not by the subsidy.

More decisively, and with an identifiable ritual component, Chen (2010) demonstrates with data from the 1997–1998 Indonesian financial crisis that religious intensity functions as an ex post social insurance. Households experiencing a $1 decline in monthly per-capita nonfood expenditures were two percentage points more likely to increase communal Koran study (*pengajian*) and one percentage point more likely to switch a child to Islamic school, but were no more likely to increase other communal activities or secular school attendance. Households that increased Koran study participation saw a 49 % decrease in the likelihood to need alms or credit four months later, while constant amount of study resulted in only a 23 % decrease, decreasing study participation in a 21 % decrease in the need for alms, and no participation in a 6 % decrease. Interestingly, this benefit is most marked for those households that were less religious before the crisis, as is the increase in the intensity of religious practice.

Chen (2010) also clearly rules out a set of alternative explanations. Mean reversion is an unlikely explanation for the difference in requiring alms, as this difference diverges during the period observerd. Putting a child into Islamic school is more expensive (and brings lower returns) compared to non-Islamic schools, so this cannot be interpreted as a strategy for reducing immediate costs. In households hit by hard times weekly work hours are increased, so Koran study does not result from increased free time. Also, other forms of distress (family deaths) did not stimulate religious intensity, and economic distress did not stimulate participation in other kinds of social insurance systems, possibly because they lack effective mechanisms of social sanctioning. Only the presence of banks and microfinance institutions mitigated this effect. In Chen's analysis, there is a causal relationship of economic distress stimulating religious intensity, which then supplies retroactive social insurance via resource sharing and job market aid. He suggests that the reason for this success is the more effective social sanctioning provided by religious groups that also encourage those who are in a better position to participate.

4.4 Collapse and Revival in a Simulated Social Network

Above I set out the hypothesis that increasing the costliness of an individual's signals would be an adaptive strategy for attracting new cooperative partners in the face of threats to survival. For the purposes of developing the costly signaling theory of religious ritual and the various alternative theories proposed to account for the connection between religion and prosocialty, there are two central points that are of interest. First, as shown above (chapter 3.2.1), the logic of costly signaling as understood in behavioral ecology does not apply to religious ritual without problems, as there is no clear reason why the signal cost (ritual behavior) should be connected to the variation in underlying quality (cooperativity). Can the dynamic of collapse and revival be accounted for without recourse to this problematic assumption? Second, there is some dispute over whether the human capacity for costly displays is an evolved adaptation or part of a more general cultural learning system. Is it necessary to assume that individuals have a needs-responsive cognitive module that causes them to signal more profusely in times of need, or does a general cultural learning system suffice?

To partially answer these concerns, I have prepared a model of collapse and revival in a social network, where agents receive benefits and pay costs relative to the number and quality of their connections and learn from their more successful partners. The key assumption here is that in an evolving social network where new relationships form while old ones degrade, costly signals have advertisement value that facilitates the formation of relationships between agents. Specifying an evolving network context for the agents resolves three issues. First, if signals are about pledging oneself to social relations, the presence/absence of these relations must be included in the model. Second, as discussed above, social harms like ostracism and reduced cooperation are threats to survival and therefore salient to the human brain (Boyer and Liénard 2006, 602). Fluctuations in network density (decreased connectivity between the agents) can thus be taken to model stochastic events that present a threat to survival. Third, cooperation can thrive in a sufficiently connected network environment without any further considerations (Ohtsuki et al. 2006; Nowak 2006). There is no need to explicitly model the interaction between cooperators and defectors, which would in this case unnecessarily complicate the model. I am thus considering only the advertisement value of the signals here. If they make the formation (and maintenance) of social relationships more likely, does a time of social need make them more adaptive?

To test whether it is necessary to assume that individuals have a needs-responsive cognitive module that causes them to signal more profusely at times of need, this feature was left out of the model specification. Agents merely have an aptitude to learn the signaling behavior from those around them. In practice, each agent will have a signaling value that is conceptualized as relative adherence to a social norm that prescibes costly behavior (such as the observation of Sabbath, which can be complied with to a varying degree). However, as individuals routinely engage in social comparison to deduce the proper content of such norms (Bicchieri 2006, 178–181), agents can alter this value based on how much their contacts are doing it. Models of social evolution (Henrich and McElreath 2003, 129–131) suggest the existence of a success bias (do what the most successful other is doing) and a conformist bias (do what most others are doing), but the former is thought to be the stronger of the two (Henrich 2004, 26), so it has been chosen for this simulation.

Model Specification

Jin, Girvan, and Newman (2001) have proposed a formal model of evolving social networks based on a minimal set of features:

1. *Fixed number of agents:* Although new agents are introduced to real-world networks from time to time while some others leave, the timescale for this is longer than making and breaking new connections, which justifies leaving this complicating factor aside.
2. *Limited degree of connections an agent can possibly have:* Degree distribution of empirical social networks does not follow a power-law distribution, but rather peaks around a specific value (which is dependent on the criteria for counting contacts). This is likely to result from the fact that there is a cost to maintaining relationships, and resources for this do not vary greatly across individuals.
3. *Clustering:* Empirical social networks have, to a high degree, the property of transitivity – friends of my friends are likely to be my friends also.
4. *Decay of friendships*; If an agent can have only a limited number of connections, the evolution of a network stagnates unless old ones are occasionally removed. This would be true of real-world networks also, as friendships grow cold,and acquintances may never be met again.

The present model employs the "Model I" specified by Jin, Girvan, and Newman (2001) for creating arbitrary social networks with the abovementioned features. In a population on N agents, on each timestep the probability of two agents i and j establishing or re-establishing a connection is $p_{ij} = f(z_i)\,f(z_j)\,g(m_{ij})$. Function f depicts the falloff in probability after the number of contacts an agent has (z_i, z_j) has reached the specified limit z^*: $f(z_i) = \frac{1}{e^{\beta(z_i - z*)} + 1}$. Function g represents the increase to base likelihood (p_0) that two agents form a connection if they have a number (m_{ij}) of mutual contacts: $g(m_{ij}) = 1 - (1 - p_0)e^{-\alpha m_{ij}}$. The decay of connections is modelled by giving each connection a strength $s_{ij} = e^{-\kappa \Delta t}$, where Δt is the number of timesteps since the two agents last met. After s_{ij} is less or equal to a threshold value θ, it is considered nonexistent for all purposes. Parameters α, β, and κ control the rate of change in these functions.

The present model modifies this specification by setting the probability of two agents meeting to $p_{ij} = f(z_i) f(z_j) g(m_{ij}) h(d_i) h(d_j)$ instead. Here function h represents the advertisement value of costly signals and has the form of a logistic curve: $h(d_i) = \frac{1}{1+e^{-\lambda(d_i-0.5)}}$, λ being an adjustable parameter.

The model is initialized into a state where there are no connections between agents, and each agent has a signaling value $d_i \in [0,1[$ with continuous uniform distribution. In each timestep, after connections are formed and degraded, agents cooperate by giving each of their contacts the benefit b, and paying the cost c times the number of contacts they have to form their payoff p for this timestep. The benefit can be weighted in various ways, but in the runs reported it is multiplied by the strength of the connection between the two agents s_{ij}, as it seems prudent to employ a property already present in the model. This does not mean that the assumption of underlying quality is introduced into the model, as the connection strength emerges from the advertisement value of the signal in reinforcing already existing connections. The costs could also be discounted correspondingly, but as one may surmise that while benefits of cooperation can easily dissipate in low-quality relationships, the costs are more "robust," so this is not done here. After this, the score for each agent is set to be $\frac{v_{t_n}}{v_{t_d}} = \frac{\omega v_{t-1_n} + p(1-d_i)}{\omega v_{t-1_d} + 1}$ (t and $t-1$ refer to this and the previous timestep, and ω is a discount parameter).[5] Finally, cultural learning is modelled by having each agent i observe their contacts (if any) and select the one with the highest score (j) to learn from (if tied, one is picked randomly, and if no contact has a greater score than the self, learning does not happen in this direction). It then sets its signal value to be $d_{i_t} = \frac{d_{i_{t-1}} + d_j}{2}$. At this point there is a fixed probability for "mutation" or "innovation" (also for those agents with no contacts), in which case d_i is set to a random value $0 \leq d_i < 1$.

5 If payoff p happens to be negative, the term $p(1-d_i)$ above is set to just p instead. Negative payoffs can happen if the average benefit from contacts is less than the cost the agent pays for them (due to low contact strength). This may not be desirable, in which case parameters should be set so that $\theta b \geq c$.

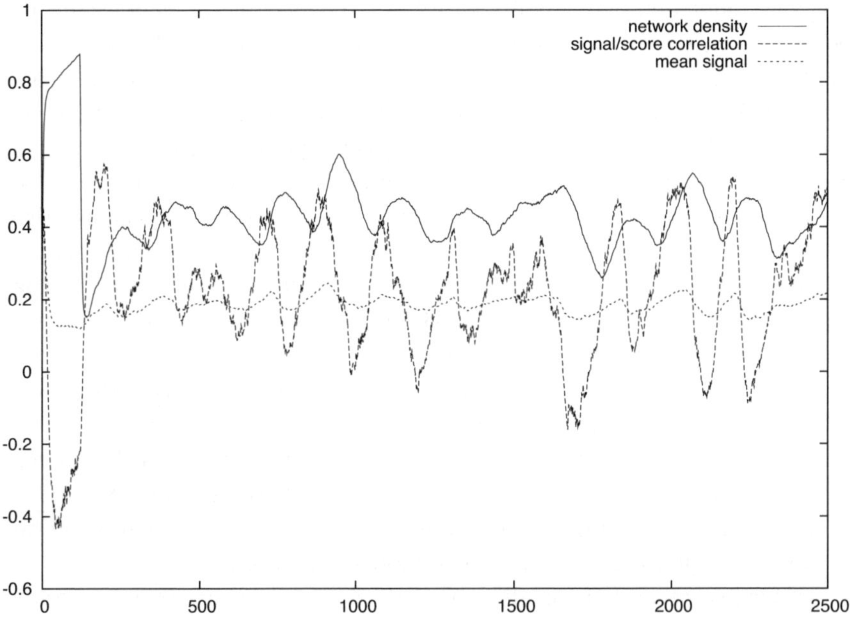

Figure 1: Example run of the simulation with parameters $N = 2053$, $\alpha = 5$, β=2, θ =0.3, $\kappa = 0.01$, λ = 10, ω =0.9, b = 4, c = 1, p_0 = 0.001, $z^\star$ = 5, and mutation rate of 0.005

Model Behavior

The model used in reported simulations was implemented on NetLogo 4.1 (Wilensky 1999), with the program code available from the author by request. A sample depiction of the overall behavior of the model can be seen in figure 1. As predicted, the adaptiveness of costly signals (calculated per timestep as the Pearson product-moment correlation coefficient between individual signal strength and score) peaks at times when there has been a drastic drop in network density, falling again as the network is reconstructed (a density of 1 means here the situation where every agent has $z^\star$ contacts). The signal mean peaks simultaneously with the adaptiveness, and drops in network density are preceded by drops in signal mean. The signal mean varies in a rather small range, but as all numeric values in the model are arbitrary, this is of little consequence; the focus is on qualitative changes in the model behavior.

Sensitivity of the dynamics to parameters was tested with runs on combinations of parameter space $\alpha \in \{2,5\}, \beta \in \{2,5\}, \lambda \in \{5,10\}$, $\kappa \in \{0.01, 0.05\}, p_0 \in \{0.001, 0.005\}$, θ = 0.3, ω = 0.9, b = 4, c = 1, $z^\star$ = 5, N =

2053 and a mutation rate of 0.005. With $\lambda = 5$, the macro-level relationships between density, signals, and score became harder to discern, with one or more graphs fluctuating too intensely, but even in these cases what can be clearly seen is a sharp drop in density following the initial formative phase of the network, at which point the adaptiveness of signaling peaks. Outside these runs, it was found that with a higher mutation rate (0.001) behavior of the model becomes more chaotic, and with a lower one (0.01) it is exceedingly stable. Also, if the population size is too low (e.g. $N = 250$), the model becomes more susceptible to random fluctuations (which was to be expected). However, with smaller populations it is possible to monitor visualizations of evolving networks and make further observations:

- Large subnetworks tend to develop towards low signal means as those with the lower signals reap higher benefits in a stable network (and others imitate them). However, this contributes to network dissolution as these low-signalers are unlikely to form new relationships to replace decaying ones. When this happens, it is those agents that happen to innovate with high signal values that form the nucleus of the replacement network.
- In a stable network, innovations to high signal values usually return to the fold quickly (as they squander their benefits and decide to follow the example of a more prosperous neighbor). But even parts of larger networks can go through periodic "revivals" if it happens to be a local leader who starts suddenly sending costlier signals, causing their contacts to imitate them. This dissipates after some time (since their relative success goes down), leaving the participants less successful but with stronger bonds between one another. The latter makes this cluster more likely to survive when other parts of the network collapse.
- Throughout the simulation there are also small isolated clusters of agents that retain high signal value. They tend to survive as such since their low number makes the likelihood for any of them innovating with a lower signal value low, their high interconnection makes it unlikely for any of them to make new outside contacts, and their high signal values make it more likely that they recontact each other and thus maintain their network.

The model gives some credence to the hypothesis that the advertisement value of signals is sufficient to create a dynamic of collapse and revival in a social network. It also shows that the existence of such a dynamic as an empirical phenomenon can result from the functioning of a general cultural learning system. It is not necessary to assume either a connection between signal intensity and underlying cooperativeness or the existence of a cognitive module specialized in the regulation of signal intensity. It does not preclude either of these, as they can plausibly fine-tune the pre-existing adaptive dynamic.

Cyclicity appears characteristic of many properties in human systems, from population cycles to societal complexity (Ibn Khaldûn 1967; Turchin and Nefedov 2009; Gavrilets, Anderson, and Turchin 2010), which makes its emergence interesting in itself. It is worth noting that the present model depends heavily on the social network model specified by Jin, Girvan, and Newman (2001); alterations are limited to the partial signal-dependence in the probability of making a contact and the addition of mechanisms for scoring and cultural transmission. I consider this a strength of the present model, as it makes it less likely that we are dealing with artifacts of the present specification and more likely that they might reflect systemic properties of social networks in general. Even the plain Jin model was suspect to some stochastic fluctuations in network density, although these are accentuated by the changes made.

4.5 Kinds of Revivals and Cultural Inertia

Both the simulation runs and the review of empirical studies indicate that various threats to survival can stimulate intensification of religious activity, although with empirical phenomena the relationship is clearly not determinative. I also wish to emphasize that contextual factors are of central importance in defining the precise form the reform or revival will take. This is most aptly illustrated by the history of medieval monasticism in Western Europe, which appears as a continuous cycle of increasing laxity and revived strictness, but one that also evolves over time and context (Lawrence 1989). Examples include the independent Benedictine houses from the sixth century on; tenth century Cluniac and Gorze reforms; late eleventh century Cistercians and Carthusians; and the mendicant friars in the thirteenth century – to name only the most significant. Initially each of these movements was as strict as any other, but over time each became more and more entangled with worldly affairs and amassed vast corporate wealth. This prompted the revivalists (who were often numerous and not in direct connection with one another) to demand a return to the apostolic way of life as embodied in the rule of St. Benedict (or in the gospels for the friars). The canonical position of these documents worked to constrain variation (so that we can recognize a continuing tradition of monasticism), while the streams of postulants acted as the selecting force between the various reformulations. But each of these cycles also brought reforms to better cope with changing times; the fragmented political map of early medieval times favored secluded localism, but Frankish unification called for more hierarchical and federal models, until the rise of towns made the whole claustral principle unwieldy. The impact of this evolving monastic institution on European history and culture can hardly be exaggerated, and as communal experiments their success is without equal (barring other religions' monastic traditions).

But picking such different examples of religious revivalism poses the question: What determines whether the experienced threats prompt an individual to signal (more profusely) commitment to his or her previous religious identity (as in Chen 2010) or whether he or she decides to transfer commitment to a new group altogether? As the above account (chapter 4) on the origins of Christianity and Islam shows, this may be a long process, but in some cases the new identity of the group is more pronounced in already in an early phase. The Ghost Dance, for example, provided a shared identity of "Indianness" that had previously not existed among the tribes, which were divided along ethnic lines (Smoak 2006, 154).

Aberle (1965, 538) took a critical feature of emerging religious movements to be "[relative] deprivation which stems from *change*, actual or anticipated" (emphasis original). It is intuitive that if substantial changes happen over a long period of time, they may not even be perceived, may feel irrelevant to an individual making short-term behavioral decisions, or may seem unstoppable. However, if the environment appears to be in a continuous flux so that individuals cannot even attempt to follow their conventional strategies of resource acquisition and social navigation, they most likely experience feelings of uncertainty, deprivation, or threat (cf. the "mazeway" of Wallace 1956, 266–267).[6] The amounts of time and change needed to produce such feelings are necessarily context-dependent, but when analyzing the impact a situation has on agents, the key factor should be the predictability of everyday experience. Aberle's (1965, 539) analysis of the Navajo in the 1940s is illustrative: "With the decline in livestock holdings came a necessary decline in certain types of behavior viewed as desirable by Navahos [sic]. Kin did not fulfill their obligation to kin, neighbors to neighbors, 'rich' to poor, because the wherewithal for reciprocity and generosity was no lon-

6 Wallace (1956) used the concept of mazeway for the mental image an individual maintains of him- or herself and the surrounding culture in order to perform stress-reducing actions. It can become incongruent with reality, and at these times a "revitalization movement" occurs, depending on the nature of the disturbance, to purge one system of problematic elements or introduce new ones into it. Wallace's model has been influential but suffers from the superorganismic, *Gestalt* view of culture and society where transitions occur from one stable state to another with the aim of conserving a specific "culture" (unless they "fail"), whereas a population-oriented view can more readily account for the intra-societal dilemmas and centrifugal tendencies of cultural change (see Richerson and Boyd 2005, 7–8).

ger there. There was a pervasive feeling that people did not *behave* as they should, or as they once did."

A similiar tale can be told regarding the Christian revivals of late eighteenth and early nineteenth century Finland. Relatively independent cases of people flocking to hear self-proclaimed prophets occurred across the country. This led to the formation of revivalist groups that practiced both emotional and costly signaling, with a sectarian identity that created tensions with more conventional believers (Heino 1976; Raittila 1976; Ylikangas 1979, 2002; Sulkunen 1983, 1999). Historians have described how during this era substantial societal changes were underway (Soininen 1980, 389–391; Troberg 1995; Karisto, Takala, and Haapola 1997, 22–24, 29, 132–136; Toivanen 1997; Sulkunen 1999, 47–48; Saarenheimo 2003). Land reparcellment acts changed the fabric of rural society, as previous methods of common farming and living were forced to give way to a more enterpreneuring, individualistic ethos. Lucrative tar and timber businesses created common pool resource dilemmas in forests that had previously been open for settlement and foraging. The fast growth in population led to increased deprivation, both with basic subsistence among the now landless population and with social status among the children of householders who now had to seek their fortune elsewhere. Old class distinctions were also ruptured when some landowning peasants had incomes equal to some members of the old aristocracy. This too would seem to be an example of a situation where individuals' expectations as to the behavior of those around them are constantly defied, a condition fatal to norm observance and cooperation. I will return to this case in more detail in chapters five and six.

To answer the question posed above, I suggest that if the adaptive challenges are perceived to be only temporary disturbances (external threat, quantitative changes in resource availability) in an otherwise stable system, amplification of existing commitments may appear viable. But if people are experiencing difficulties of a more qualitative nature in predicting what is happening in their surroundings, such as new internal divisions or disappearing resources, they may also anticipate that old commitments may no longer be valid. Writing on the formation of Millenarian movements in the Middle Ages, Cohn ([1970] 2004, 87) notes:

> Poverty, wars and local famines were so much a part of normal life that they were taken for granted and could therefore be faced in a sober and realistic manner. But when a situation arose which was not only menacing but went altogether outside the normal run of experience, when people were confronted with hazards which were all the more frightening because they were unfamiliar – at such times a collective flight into the world of demonological phantasies could occur very easily.

In this case they have occasion to identify other individuals in a similiar predicament. This is clearly a signaling problem and may contain a clue as to why religious groups tend to cluster around socioeconomic variables. At any single place and time, it is more likely that individuals who are similiar to each other in one or more ways experience hazards to their survival.[7] This is a tentative suggestion on my part, but it echoes Glock's (1964) contention that new religious movements form when individuals become aware of being in an unspecified threatening situation. Incidentally, Ylikangas (1979, 2002) also finds in his study of the Ostrobothnian Revival that the emerging, relatively "deprived" middle class of householders, town craftsmen, and lower clergy joined their ranks into a *religious* protest because they were unsure as to what they were protesting against.

But critics of deprivation theory questioned why people remain in sects and churches even when the postulated deprivation seems long gone. If we posit formation of religious institutions as a response to survival hazards, why is there considerable cultural inertia in their "powering down"? There are several factors that may each contribute to this. First, if one finds his/her needs in life being satisfied, how does one go about determining how this happy state of affairs came about? Especially if there has previously been a state of want which is now gone, it may be safer to let things remain as they are. Reviewing cases of institutions that evolved in managing common pool resource dilemmas, Ostrom (1990, 208–209) concluded that as long as circumstances remain good, institutional changes are unlikely to be made, but a series of bad years leads to amendments. Potential losses are weighted more heavily than potential gains (in short, the wisdom of "if it is not broken, don't fix it" applies here). However, the situation would seem to be different for later generations who have not experienced these hazards (Ibn Khaldûn 1967, 105–106, 109, 123). As mentioned above, Norris and Inglehart (2004, 76–78) show that the frequency of religious participation can plausibly result from the experience of existential security in the formative years. If insecurity breeds religiosity, those who grew up in highly secure environments (such as European welfare states) may have difficulties in understanding their parents' and grandparents' attachment to old religious institutions (although even some contemporary Finnish youngsters view the state church as an "insurance corporation" they remain with for fear of other support structures collapsing; Wilska 2006, 105–106).

---

7 Of course, if the hazard is not a widely shared experience, there are no grounds for group formation, and the scattered individuals are left to cope with it as they can. Also, when social commitments are being renegotiated, emergence of reciprocity is less likely to be reliant on punishment, but rather on displays that signal trustworthiness and commitment.

Second, when a norm or system of norms becomes established, they may remain in place even when they become disliked by the majority (Bicchieri 2006, 179–188). As outlined above, social norms rest on empirical and normative expectations; if I believe a sufficient number of others to abide by a norm, and they expect me to conform to it, I have reason to believe the norm is in place. If everyone infers the existence of a norm from the behavior of others, and consequently abides by it, even an unpopular norm may remain in place. Furthermore, individuals are apt to mis-infer others' behavior to result from their preference for it, even while they acknowledge the role of social pressure in determining their own behavior. This may result in a vocal minority upholding a general conception of unanimosity from which individuals are afraid to deviate for fear of social stigma and punishment. Here the costly, easy-to-monitor aspects of religious norms come to the fore, as it will be more difficult to quietly defect if norms dictate the minutiae of daily behavior, and every departure can be viewed as implying defection.

Third, cognitive scientists of religion have made a case for the qualities of religious beliefs that enhance memorability and ensnare our cognitive faculties. Representations like immortal humans or talking images are minimally counterintuitive (contain a single breach of our intuitive expectations) and are remembered better than those that are fully intuitive (Boyer 2001, chapter 2; Atran and Norenzayan 2004, 721–724; Pyysiäinen 2009, 22–28). Rituals grab our attention with either high emotional and sensory arousal or frequent repetition (Whitehouse 2000; McCauley and Lawson 2002) and with implications of the danger of nonperformance (Boyer 2001, chapter 7; Liénard and Boyer 2006). Regardless of possible societal functionality, these effects enhance the staying power of institutions underpinned by such representations via a process of cultural selection (cf. Sperber 1996). And whereas the two other factors outlined above apply for the most part to both religious and nonreligious institutions, these features are more emblematic of the former.

4.6 Chapter Conclusions

In this chapter I have attempted to take the empirical-theoretical synthesis of chapters two and three a step further by building an argument of religious revivalism being a functional response to experienced deprivations and hazards. Of existing models, the deprivation theory has been found lacking, as it has been unable to account for the increase in members' social status and the fact that non-deprived individuals are in no hurry to leave their groups. As I argued, the former can be explained by the benefits brought by the formation of a cohesive fellowship, and the latter by the tendency of groups to become locked into practicing their norms regardless of their continued relevance. A more recent model of a hazard precaution system addresses ritualized behavior on the individual level only and grants ritual *per se* little adaptive function. However, I argue that when viewed in the context of findings on ritually induced cohesion, these behaviors appear more meaningful. This view receives support from a number of correlative studies from different cultures that show a connection between environmental threats and the prevalence of religiosity or revivals. That this kind of dynamic can result from signaling behavior on the individual level is also supported by the simulation experiment.

An interesting theoretical question is what determines whether the upsurge of religious activity takes on a separatistic (or otherwise identity-altering) nature. I argue that a key factor in analysis should be the predictability of individual experience – whether an individual is able to anticipate how the social and material environments function and navigate these successfully. The more drastic and qualitative the disruptions to existing systems, the more likely it should be that individuals seek new commitments to supplant the old ones and have occassion to engage in some kind of signaling activity to identify others with similiar needs. The present model does not assume a linear evolution from sect to church and is more sensitive to contextual factors, both in determining the need for strong norm conformance and the conditions in which to expect gradual relaxation unless constrained.

In the next two chapters, I explore the theoretical ideas of this and preceding chapters in the context of a single case, namely the Christian revivals that occurred in late eighteenth and early nineteenth century Finland. They appear to confirm and elaborate many of the predictions I and other authors have sketched, but they also illustrate the negotiatory character of these group processes, which makes them hard to capture within theoretical formulations.

5 Revivalism in early Nineteenth Century Finland

> The influence of religion, then, decreased among men, and they came to use restraining laws. The religious law became a branch of learning and a craft to be acquired through instruction and education. People turned to sedentary life and assumed the character trait of submissiveness to law. This led to a decrease in their fortitude. (Ibn Khaldûn 1967, 96)

5.1 Revivalism as Homogeneous Phenomena

The period of mass revivals in Finland, then part of Sweden, begins in the early eighteenth century, after the Russian occupation of 1713–1721 (*isoviha*). Some of these can be taken as examples of international influence, such as the spread of Herrnhutism that took hold among the literate folk of coastal towns after 1734, and the so-called "Estate Pietism" (*kartanopietismi*), a short-lived upper class phenomenon based on ideas brought by war prisoners returning after the Great Northern War (Ruutu 1967, 95; Ylikangas 1979, 19–20).[1] Some of them were more radical: the notable mystic-separatist community of former petty officers Jakob and Erik Eriksson in 1720s Ostrobothnia that forbade all interaction with outsiders, church communion, and infant baptism, and met its demise only after the brothers' banishment and *pikkuviha*, the new Russian war in 1741–43 (Ruutu 1967, 97; Ylikangas 1979, 19; Sulkunen 1983, 3). But in the latter half of the eighteenth century the whole country seems to have burst with the ecstasy of revivalism, so that by the end of the century local revivals had appeared all over the country, excepting only the northernmost regions (Airas 1936, 256–257). These revivals were ecstatic in nature, involving trances, glossolalia, visions, preaching asleep, and with women and children especially prominent (Ruutu 1967, 106; Ylikangas 1979, 20), although with the passing of time men increasingly replaced them (Sulkunen 1999, 82). Also, there is little evidence of contacts between these locally occurring ecstatic folk revivals and international currents (Airas 1936, 256; Sulkunen 1983, 5). Rather there was "a spontaneous readiness for revivalism from the 1770s onwards all over Finland" (Siltala 1992, 27).

Historians usually distinguish between the eighteenth century "folk revival" and the nineteenth century revivals that resulted in the creation of unified revival movements (the Supplicationists, the *Skörts*, the Laestadians and the Evangel-

---

1 On the other hand, the anti-church protest of Lars Ulstadius, Petrus Schäfer, and Olaus Ulhegius known as "Radical Pietism" (*yltiöpietismi)* carries an ill-fitting name, as the inchoate German Pietism had not yet even reached the German provinces held by the Swedish crown in 1689, the time they were imprisoned (Ylikangas 1979, 13).

ical movement), but apart from this, the existence of a qualitative difference is questionable. Ruutu (1967, 100, 103) considers the eighteenth century revival to have been a top-down process (in contrast to the nineteenth), claiming that the "stabilization" of southwestern revivals was due to the influence of Abraham Achrenius, vicar of Nousiainen, and that elsewhere too their continuance was dependent on clergy or "mature laymen." This is in contrast to the view presented by Sulkunen (1999, 22, 71), who finds the impetus for the revivals in ordinary folk and emphasizes the role of women ecstatics in guiding them. She sees little or no difference between the two phases of revivals but also subscribes to the oft-mentioned point (e.g., Heino 1976, 157–158) that the early revivals would have been less hostile towards inaugurated clergy than the later ones. This may well be, but it appears somewhat determined by local conditions and the attitudes of the clergy themselves. Heino mentions nineteenth century Rauma, where the Jumpers were not persecuted and which was free of anti-clergy separatism, whereas Toivanen (1997, 78–89) describes how also in earlier times an unsympathetic vicar could see the revival get out of his control.

Reading the history of revivals in the eighteenth and nineteenth centuries reveals a multitude of little-known local preachers and movements that may or may not have influenced later and better-known ones. In addition to the well-known revivals of southwestern Finland (starting in 1756 with the visions and preaching of Lisa Eriksdotter, and followed by those of Anna Lagerblad, Juliana Söderborg, and Anna Rogel in the following decades), Sulkunen (1999, 17–19, 66–82) names two prophesying women from Kurikka (1774), a men-and-women led revival in Orivesi (1775), and the revival led by tailor Fredrik Axberg in Tavastia and Nyland in 1770s. Heino (1976, 38, 67, 82) also mentions revivals at Nastola (1793), Iitti (1825), Lapväärti (1828), and the Khlysti-influenced Karelian Jumpers of the early nineteenth century (see Puolanne 1936 on the latter). Airas (1936, 256) states there is no way of connecting the "Wiklundian" Revival of 1770s Torne Valley (again instigated by young girls) to any of the southern revivals, not to mention more faraway influences, and is more willing to term it a "God's miracle." In 1798 glazier Jakob Wallenberg started his own revival at Alahärmä, Ostrobothnia (more on which in chapter 5.3.2), and while his preaching was a mélange of prior local mystics' ideas and communiques from Stockholm (including Jacob Böhme's influential *Psychologia vera*), it was peculiar to the extent of containing a free-love aspect (Ylikangas 1989, 158; Fräntilä 2001, 186–190). Later, when the more unified revival movements, especially the Savonian Revival, begun to spread in the nineteenth century, they encountered local revivals that refused to be incorporated, such as the sect of "prayers" in 1830s Lapua (Rosendal 1905, 155–156) or the "movement of friends" (*ystäväliike*) led by Margareta Högman and Salomon Häkkänen (Koitti 1976; Hökkä 2006). Sulkunen

(1983, 5–6) sees it as a distinct possibility that these revivals, however similiar they were to each other, emerged independently first in the coastal regions and later in inner and northern parts of the country (for a map see Ruutu 1967, 96). Thus, in a hundred years' time, the whole of Finland was taken up in revivals.

The listing above is not intended to be exhaustive but rather to illustrate the number of little-known revivals, for which there seemed to be a sudden "demand" or "drive." Also, only a portion of these proto-movements survived, most apparently either dying out (some from persecution) or being assimilated into others. In comparative perspective this is not surprising. Reviewing the available data on nineteenth and twentieth century American sects, Stark and Bainbridge (1985, 126–134) found that "nearly a third of all sects (32 percent) reached their high-water mark on the day they began," either declining or stagnating after that, while only 11 percent had seen fast growth even momentarily (naturally, these figures do not include those groups that lived and disappeared without a trace). There is also moderate support for a claim that the number of adherents to religious groups follows a power law distribution (Clauset, Shalizi, and Newman 2009),[2] which would mean that the overwhelming majority of groups that have existed (and will exist) are negligibly small, and only few reach large or even intermediate sizes (i.e., become ethnic, national, or world religions). Applied to the Finnish situation, this would mean that in addition to those few large revival movements that grew to national significance, and a somewhat greater number of local revivals preserved in historical records, there were numerous incidents of prophesying that did not take hold, proto-groups that dissolved because of internal disputes, silent intervention by local authorities, or were peacefully assimilated to other, expanding movements.

Hartikainen (2005, 275) remarks that scholars with a background in one of the revival movements have been apt to over-emphasize that movement's distinctiveness from any other. Conversely, others have noticed that except for some minor differences in doctrine and practice, for members of various revivals the new way of life held a shared significance. A line of evidence that substantially favors the latter interpretation can be found in the observation that as a new wave of revivalism spread from some point of origin, it often bypassed

2 Quantity x is said to obey a power law if it follows a probability distribution $p(x) \propto x^{-\alpha}$, where α is the scaling parameter typically in the range]2,3[, and there is often also a minimum value x_{min} above which the law will apply (Clauset, Shalizi, and Newman 2009, 662). Scientific interest in power law distributions results from the fact that many natural phenomena seem to follow it, and its occurrence hints at there being underlying systemic properties (such as preferential attachment) that generate it.

those regions already taken up by another vigorous revival (Sulkunen 1999, 79). To pick an example, Lohi (1997, 657–658), in his careful exposition of the geographical expansion of the Laestadian Revival in the latter half of the nineteenth century, notes that the movement filled the gaps left in the national map by other revival movements, as these movements most often formed an "impenetrable wall" the Laestadian message could not breach.

According to a traditional view of church history, the revivals were a result of Pietist trends coming from abroad,[3] which took hold in Finland and were transformed into a protest against the stale and shallow devotional life offered by the state church orthodoxy (Salomies 1962, 369–410). By reading translations from the English puritan Arthur Dent, seventeenth century Baptist Thomas Wilcox, German shoemaker and mystic Jacob Böhme, among others, Finnish Christians could connect to the intellectual climate of Protestant Europe without intervention from the clergy or the gentry (Ruutu 1967, 103–105). Ylikangas (1990a, 320; also 1985; 1991; 2002, 10–11) has been vocal in his criticism of such ideohistorical positions, which (he claims) are often employed in theological ("confessionally involved") research and assume that a new faith reaching our shores awakens a spiritual need in individuals encountering it, without any determining structural factors. Were it so, he argues, there should be no difference between the social profile of the movements and that of the general populace, and their regional distribution should follow only population density and route networks – but this prediction does not hold (see chapter 6.1). Nor were the ideas that were taken up always in tune with the latest trends of Protestant Europe. The Jumpers of Satakunta were known to read the writings of early seventeenth century Catholic mystic Jean de Bernières-Louvigny (Heino 1976, 122), and Luukkanen (2001) describes how the Ostrobothnian mystics incorporated into their texts some writings of the Fransiscan Aegidio (Blessed Giles of Assisi), that had taken some 550 years to travel from Italy to Finland! Regardless, early nineteenth century folk revivalists considered these current and topical.

So, instead of trying trace the spread of religious ideas among a "religious innovation network," we have to look at the societal structures from which the revivals may have emerged (Sulkunen 1983, 5–7). But while this criticism is certainly on the mark with such ideohistorical positions as outlined above, I see less reason to construe this as a unified "theological" research agenda, considering the views put forth by Juva (1962) and Heino (1977). Juva's (1962) brief ex-

---

3 It should be noted that the seeds of such a view where already contained in the state authorities' desire to stop the conventicle placard of 1726, which they considered to be dangerous foreign currents disrupting the social order (see ch. 5.2.3).

position gives due weight to societal change as an *explanans* for religious change. Heino (1977) argues (in a way similiar to Ylikangas 1990a) both for consideration of social context and against the genealogical premise based on superficial similarities between movements. However, he also notes that it is misguided to "explain away" the pervasive religious dimension of revival movements, as attempts to treat them like any other social movement (as has been done in more sociologically inclined studies) lead to confusing and unsatisfactory theories. This said, the sociological and sociohistorical explanations offered by Ylikangas (1979, 1990a) are compelling, and the scant attention they have received since (also noted by Huhta 2007, 33) is curious. With this in mind, I am reviewing his findings in conjunction with other accumulated research, while at the same time attempting to demonstrate how the signaling hypothesis circumvents the abovementioned problem of structural explanations by giving religious expression a privileged position as a motivator for social action.

5.2 The Socioeconomic Context

In this subchapter I provide a background on the socioeconomic situation in Finland from 1750–1850. Based on their prominence in the revival movements, the focus is on changes in the operational environment of householders and craftsmen, although I situate this in the context of wide-ranging societal developments. An important element of Finnish society during this period was the four estates: nobility, clergy, burghers, and householders. It was through them that people of the realm were represented in parliamentary sessions. But this division excluded a large portion of the populace, most notably the landless agrarian folk, and of clergy only vicars had the vote. Even though the specific privileges granted to each estate were subject to criticism and change, this division in itself was seen as God-given and untouchable (Wirilander 1974, 15–22, 29).

Until the "Finnish war" of 1809 (*Suomen sota*), Finland was part of the Swedish realm, subject to the Swedish crown and its laws. After this Finland became (until 1917) an autonomous grand duchy subject to the Russian czar via his representative, the governor-general. The Finnish were, however, able to retain their existing legal system, languages, and religion. For this reason the 1809 change in the topmost levels of power hierarchy does not appear as a substantial disruption in the developments during the observed period. To the contrary, during the fifty years that followed the war, Finland was in some respects stopped in time, as no substantial legal reforms were made (Myllyntaus and Åström 1980, 334–335). The crown and its chancelleries were replaced by a senate chaired by the governor-general, and the diet of the four estates retained its parliamentary role. But as there were no assemblies between 1809 and 1863, governance of the country rested solely upon civil servants.

5.2.1 Land Reforms and Growth of the Agrarian Populace

During the latter half of the eighteenth century, the population of Finland grew rapidly even by European standards, doubling from half a million to about one million (Koskinen et al. 2007, 59–60).[4] The cultivated area grew in proportion with the populace and could adequately sustain it, but the difference between good and bad years could be steep, as were differences between population groups, with farmers selling their grain in towns while locals went hungry (Soininen 1974, 135–138, 351–356, 366–370).[5] Parts of the agrarian populace experienced social decline, and there was not enough land for all children of landowning farmers, despite the splitting of farms and new settlement. They became part of the growing landless populace, a group struggling with diverse means of finding subsistence and a continuous threat of seasonal unemployment (Koskinen et al. 2007, 65). According to Soininen (1974, 302–303, 382–389), their possibilities were few. Fishing rights were often bound to land ownership, and resources required for peasant sailing or horse freight were such that only householders could afford them. Agrarian handicraft was regulated so that for most, only cottage industry was allowed. Also the slash-and-burn agriculture practiced in Eastern Finland (Savonia and Karelia) underwent a severe crisis (from increased population pressure and over-exploitation) during the second quarter of the nineteenth century, but also here it was the householders who had better means of adapting than the landless populace.

4 Between 1750 and 1850 yearly population growth was 1.2 % in Finland, while in Sweden it was 0.7 %, in England 1.1 %, and in France 0.4 %. The high number for Finland results mainly from the eighteenth century growth. See Koskinen et al. (2007, 60) for more details.

5 The most dramatic example of this occurred with the Great Famine of 1867–68 (Soininen 1974, 403–404, 408). During this single year, the population of Finland declined by 100,000 souls. The period of 1862–67 had already seen several crop failures. Most of the agrarian populace had no means of purchasing imported grain. When food ran out, people left their homes to seek out those who still had something left. The masses of beggars included not only the landless but also poorer peasants. Their number is estimated to have been in the tens of thousands by early spring 1868, with tens of beggars visiting large houses of southern Finland every day. They were continuously on the move, as any single location was quickly depleted, leading to further suffering and the spread of diseases such as typhoid. Religious activity was also on the rise during these years (Häkkinen et al. 1991, 236), and the famine seems to have boosted the spread of the Laestadian Revival (Raittila 1976, 149; Lohi 1997, 24).

During the next century these differences became even steeper (Koskinen et al. 2007, 65; Soininen 1974, 370). The tax rolls of 1800 reveal land to be the most significant form of wealth (Jutikkala 1949, 181, 185, 192). This meant that the significance of traditional estates diminished among landowners, as those who owned large swathes of land became a breed apart, regardless of whether they were gentry or not (Soininen 1980, 389). One's ancestry ceased to predict one's affluence or leverage. The single most significant reason for these changes was *isojako*, the enclosure act of 1757 (Saarenheimo 2003, 349–354). It stated that should even one householder demand it, village lands must be reparcelled in order to leave behind methods of common farming and give each farmer a single, private block of land. The state authorities wanted to promote individual enterprise and progress in farming methods. In practice, requests for this were rare, so (as decreed by the act) it was often forced during taxations despite active or passive resistance. Agrarian subsistence was fragile, and change spelled insecurity. Most hindrance was caused by latent boundary disputes that could take decades to resolve. The division of forests was especially problematic, as was the basis for equitable decisions.[6]

Although its exact consequences are hard to disentangle, the socioeconomic significance of this reform was enormous, as it altered the basis of agrarian ownership. Every plot of land now had a named owner, and more punctilious maps ended the continuous disputes about boundaries (Saarenheimo 2003, 356). This pitted the landowning and landless classes against each other, as the former could now maximize their profits, while the latter (whose proportion grew quickly) had less and less control over their lives (Karisto, Takala, and Haapola 1997, 22–24, 29). Previously the landless had had access to commonly owned forests, which had now become the property of the crown and householders. The landless could no longer live, graze, hunt, or gather wood there rent-free. To the degree that the aim of creating single blocks of land succeeded, it also pulled apart the traditional village communities, in which people had lived in close proximity to each other.[7] (Saarenheimo 2003, 356, 361, 364) This made traditional forms of

---

6 For a description of the wrangling enclosures caused in a single parish, see Toivanen (1997). The process took the whole of the nineteenth century to complete, with only parts of Northern Finland left untouched by the end of it. Detailed studies regarding local processes are yet to be made, but it can be said that it was mostly in Tavastia and Southwestern Finland that communal farming continued to a greater extent (Saarenheimo 2003, 355–357). In 1848 another act (*uusjako*) was given to further reparcel the land, which more decisively split old village communities in Southern and Western Finland (Korhonen 2003, 461–463).

7 As leaving old home and community behind was for many great a change, it was often forced; only later did it result naturally from the then-altered conditions (Saarenheimo 2003, 360–361).

cooperation and solidarity impracticable, despite the continuing need for them (Karisto, Takala, and Haapola 1997, 132–134). Requests for collective efforts might receive the reply, "Everybody's got their own farm now" (Toivanen 1997, 52–53). Sundberg (1993, 164–168) outlines very similiar processes in contemporary Sweden, and concludes that the increasing competition for available resources made social relations all the more decisive in their usage. Also the interdependence of these communities had worked as a levelling force, and without it income differences started to grow, as large houses could now rent their forests to crofters. Earlier crofters had been on almost equal footing with householders, as labor shortage kept rents down, and forestry as a trade was undeveloped. With the growth of the landless workforce and tar and lumber businesses, the gap widened (Soininen 1974, 370–372, 398).

The enclosure act was not the only disrupture to the traditional establishment. According to Jutikkala (1958, 222–230, 280–283), during the eighteenth century the householder's hold on his land or house had been tenuous. In many cases members of higher estates could assume the ownership of a farmstead when the occupant died or tried to sell it. The crown was interested mainly in exacting revenues from this trade, and the predominant view was that peasants should not own land lest they grow unindustrious due to lack of outside control. The proprietary rights of peasants were simply not what would today be understood as "ownership" – whether their lands were nominally hereditary or on rent from the crown. Nor did mercantilistic control allow them to hire outside help according to their need, and the act of 1739 limited even the use of peasant's own children as workforce to one son and one daughter. This was not done out of any concern for child welfare but to guarantee sufficient workforce for manors, vicarages, and mills.[8]

A major change to the former state of affairs was the Act of Union and Security (*Förenings- och säkerhetsakten*; *yhdistys- ja vakuuskirja*) in 1789. In accordance with ascendant physiocratic ideals, it granted the householders many of the legal, political, and economic rights held by other estates. Instead of having to trade with city burghers they could transact business with local craftsmen and

---

8 According to Jutikkala (1958, 218–220, 280–283), the householders, even though one of the four estates, were also excluded from political power; for example, they did not have membership in the "secret committee" (*sekreta utskottet*; *salainen valiokunta*) responsible for state finances and foreign policy. Their weight diminished even in local government with centralization of power and the need for civil servants such as judges to be properly trained instead of being selected from among the local populace. Despite this, they were burdened with a number of obligations, such as maintaining public roads and buildings, transport of mail, and government officials.

each other, keep hirelings as they wished, and most importantly, have the same rights to their hereditary estates as the gentry.[9] Also those whose farms were on rent from the crown now had a more secure hold on them, and it again became possible for them to purchase a hereditary status. This option became very popular, first in the western provinces and then during the early nineteenth century in other parts of the country. Jutikkala (1958, 293–301; also Kuisma 2006, 131–135) considers the act to have paved the way for the future, to a "bourgeois" society with no distinctions in estate or in land ownership, and freed the householders from the state authorities' grip. But in the late eighteenth century it did not necessarily appear as such. Kuisma (2006, 142–143) states that despite all these reforms, when common forests where partitioned and the crown took what it considered to be "in excess" from the householders, while allowing the gentry to keep theirs *in toto*, it made plain to everybody where societal power resided.

Expansion in world trade and depletion of forests in Western Europe had created a demand for tar, which by the eighteenth century had become the main export from Finland. Forced by money-based taxation, peasants in Eastern Finland and Ostrobothnia kept the tar-pits burning on credit supplied by merchants from coastal towns, becoming indebted and thus dependent clients in the process (Kuisma 2006, 31–37). Prior to the establishment of Finland's first banks in the 1860s, it was trading houses that handled functions such as giving credit and taking deposits. Only as householders became more prosperous were they "freed" from their patrons (Åström et al. 1980, 438; Ranta and Åström 1980, 287–289), and after the mid-nineteenth century, householders increasingly financed each other while the poorer folk continued to either live in debt to the burghers or have no occasion for such investments (Markkanen 1977, 80, 205).[10] But depletion of forests was becoming a problem, pushing the tar heartland northwards. Collective ownership of forests had made it rational for everyone to cut as much forest as they could, without much concern for regrowth; if they didn't, somebody else was going to cut it anyway. When land reparcelling

---

9 Hereditary householders no longer had to offer their farm to the crown before selling it outside the family, nor could they lose it to the crown because of three years' unpaid taxes. The authorities also lost their right to inspect how the householder managed his farm, and he now had full hunting rights on his own lands. Some restrictions remained regarding forestry and splitting of farms. See Jutikkala (1958, 294).

10 One must not consider householders as simple victims who were exploited in this system. They themselves defended it (Ranta and Åström 1980, 288), and appear in some sources as quite capable of weighing the various economical aspects of their options (Kauranen 1999, 147–149).

got underway, such overuse escalated, ending only (with what was left) after each plot was privately owned.[11] (Kuisma 2006, 31–37, 48–49, 140–141)

To summarize, while during early eighteenth century the position of the agrarian populace had been fairly uniform regardless of land-ownership status, and the householder estate was separated from the three other estates by a divide alongside the rest of the populace, the legal developments later in the century altered this environment so that the householders now rose in rank to be at least on par with the other estates, whereas the divide now lay between them and the landless people. Every factor in this development brought the landed group upwards and the landless group downwards, according to Jutikkala (1958, 341).[12] It can be said that the established cues of social prestige were fast becoming obsolete, while traditional cooperative bonds were severed by legal reforms that forced people to abandon their collective ways. In chapter 4.5 I predicted such situations would cause people to seek novel commitments, and there may in fact be a rather close relationship between the revivals and the enclosure processes (see chapter 7.2.4).

5.2.2 Rigid Economic System under Mercantilism

As mentioned above, in the eighteenth century the economic system in the Swedish realm was still largely mercantilisic (Ranta, Åström, and Myllyntaus 1980; Kuisma 2006, 35–37, 88, 113). The state attempted to compile funds by controlling trade and industry, to which import and free competition were seen as detrimental. This was achieved by granting individuals and corporations special privileges while excluding other operators from their markets. Only towns with "staple rights" (*stapelstäder*; *tapulikaupungit*) could trade with foreign ports. Commerce outside towns (*landsköp*; *maakauppa*) was forbidden, and from the Finnish Bothnian coast it was permissible to sail only to Turku or Stockholm, from which actual export would proceed. This system benefited the burghers of these towns at the expense of other populace, and by mid-century there

---

11 Kuusi (1933, 12–13) describes how in 1758 reparcelling could not have been initiated in parts of Ostrobothnia for the lack of surveyors, and the issue became dire as it was observed how some landowners neglected the care (fertilization and drainage) of their more outlying lands, expecting to lose them anyway. At the same time the tar industry became exceptionally active in those forests that were still unpartitioned.

12 During the late nineteenth and early twentieth century this developed into a major issue concerning the rights of crofters (tenant farmers for manors, vicarages, and householders). See Peltonen (1992).

was widespread movement that included even Karelian and Savonian commoners acquiring export rights for Ostrobothnian towns (this succeeded in 1766, but it was only by 1830 that all the coastal towns had received full rights). Domestic sailing was also liberated somewhat, so peasants could sail by themselves to sell their products in Stockholm. The change of regime in 1809 spelled little changes to this (Myllyntaus and Åström 1980, 333–338). To the contrary, it stagnated economic reforms, as the Senate continued to strictly apply the Swedish mercantilistic laws, and considered suggestions such as relaxation of guild control and liberation of trade outside towns unconstitutional. As the Diet of Estates was not called to convene until 1863, no wide-ranging reforms took place in this period. Mercantilistic control continued to be strong, and Stockholm remained the main port until the 1840s.

In principle, throughout the period specifically ordained markets were the only legal means of trading outside towns, although demands for liberalization of trade were frequent and some development in this direction did happen. For example, trading grain was allowed in 1775 (Ranta and Åström 1980, 282–283, 292; Kuisma 2006, 96–117). But the towns retained their privileged role into the nineteenth century, forcing the agrarian populace to travel there with their goods (Åström et al. 1980, 444–445). Also, control of trade required control of craftmanship, and since in the countryside this couldn't be achieved, it had to end; professional practice of crafts was for townspeople only. But as householders had a right to craft items for their own use, controlling this was difficult. So, while the years 1604–1789 saw a number of acts that limited the extent and kinds of craftmanship outside towns (see Ranta 1978, 101, 107, 189, 282; Schybergson 1980, 428; Virrankoski and Åström 1980, 242), in practice this depended on the county governors (*landshövdingar*; *maaherrat*) who granted their charters. Ranta (1978, 87–91, 94, 99) states that in the eighteenth century they took considerable liberty in interpreting the law, so that for the most part it was a needs-based decision of the local assizes (*häradsrätt*; *käräjät*) to which the county governor bowed. This was however a lengthy process, and created frustrations later in the century when more people (especially younger sons of householders) turned to crafts for their living.

According to Ranta (1978, 114–147, 182–192), while the mercantilistic guild system dictated that even a country craftsman had to have a predefined area he could serve, and that people in this area had to seek his services, in practice this relationship was more complicated. Sometimes there was more than one practitioner of a particular trade in the same area, which led to animosity. For example, in the 1780s Kaarle Kalling, a glazier of Kalajoki parish accused his colleague of illegal trading. But a few years later the newcomer was made into a parish glazier, perhaps because Kalling (who made every effort to prevent this) was not liked much by the local people. It seems glaziers often complained of competition and difficulties in livelihood, but the local assizes did not usually heed such views and wanted more competition. Only if the assizes were themselves divided could a unified front of the senior craftsmen have an influence.

In towns, craftsmen were both more varied and prosperous, and they were organized into guilds that controlled the number of practitioners and the quality of their work (Virrankoski and Åström 1980, 242–243). In practice this meant protecting established masters from excessive competition by preventing the country craftsmen from serving the town populace (Ranta 1978, 54). The guilds were in turn controlled by the act of 1720 that forbade them from workers other than apprentices or journeymen and from selling products wholesale to third parties – only direct trade with customers was allowed (Schybergson 1980, 427–428). Handicraft and trade remained in the hands of corporations until the mid-nineteenth century,[13] and while there was a number of outside practitioners (illegal and exempted), these lacked the special privileges and security provided by the mercantilistic system (Vainio-Korhonen 1998, 22, 44–45, 82, 111–148, 189; 2002, 61–62). The established masters had a tight and protective network that made it difficult for outsiders to enter their trade if they were not born into the estate, didn't have godparents among them, or had not lived in a master's household and internalized the norms that upheld the system. Becoming a master was rare, and required not only skill but also a record of proper behavior and good relations established over the years. In these matters, social relations played a decisive role – material requirements of the trade less so.[14]

To summarize, craftsmanship offered a pathway to a better socioeconomic position. In the countryside they served the householder population, in towns the gentry and burghers. Their influence was greater than their number (Ranta

---

13 It appears there was some variation in the power of guilds over time and place. Vainio-Korhonen (1998, 142) states that in Turku they were strongest before late eighteenth century, but according to Ranta (1978, 54–55) in Ostrobothnian towns there were not enough masters for guilds to be founded (minimum three) before the mid-1700s. In their absence either guilds from other towns or the magistrate would control these affairs.

14 Regarding more industrialized manufacturing, it appears that the established agents (who were burghers, civil servants, and nobility) were often quite successful in guarding their markets from new competitors, and had the state policy on their side in guaranteeing availability of cheap resources – often to the detriment of householders (Virrankoski and Åström 1980, 251; Kuisma 2006, 88, 129–131). However, early in the nineteenth century, handicraft still dominated over manufacturing (Schybergson 1980, 408–409). By 1815 the number of industrial workers was 1,960, compared to 3,110 craftsmen in towns and 4,220 in the countryside. The situation changed only after the 1840s.

1978, 285–286). However, many of them were part-time farmers, and where becoming a crofter was an option fewer people sought employment in crafts (Ranta 1978, 283; Virrankoski and Åström 1980, 243; Schybergson 1980, 429). But if we leave the issues of sufficient skill aside, it appears that in both town and countryside, from an individual practitioner's point of view, the legal decrees were not the principal obstacle to making a career as a craftsman. Rather, what determined the issue was how well the individual could draw on his network of supporters, whether by securing a backing from the assizes or acceptance by the guild. It is perhaps not surprising to find that a tradesman's success was dependent on social relations, but in the age of mercantilism it was doubly so.

5.2.3 Clergy and the Conventicle Placard

The clerical estate was not a uniform group, and by the end of the eighteenth century growth in the number of deacons (*pastorsadjunkter*; *apupapit*) had created an internal hierarchy, with wealthy vicars (*kyrkoherdar*; *kirkkoherrat*) at the top. Young clergymen had very real problems of livelihood. From 1830–1850 there was such an oversupply of clergymen that new ordinations were occassionally put on hold (Björklund 1939, 122–123, 214). Those who were not charismatic preachers had difficulty in securing a permanent position and had to transfer from one deacon's post to another. Meager pay meant getting into debt, especially if they had families, and some took up growing staples to ease their position (Peltonen 2006, 105–106). But while those at the lower levels struggled to make a living, vicars could be quite wealthy and treated their subordinates as servants (Suolahti 1919, 186–198).[15] Vicarages were entitled to statute labor and taxes and sometimes even functioned as centers for barter, while for deacons and chaplains (*kapellaner*; *kappalaiset*), the service fees they received after officiating at a baptism, wedding, or funeral made up a substantial proportion of their earned income (Suolahti 1919, 155–174, 203; Heikkilä and Seppo 1987, 76–77; Peltonen 2006, 103).[16] But for some vicars a luxurious lifestyle proved too expensive, as offering lavish hospitality to the gentry left them in debt; there are several exam-

---

15 Suolahti (1919, 195) mentions a 1777 dean's inspection were the local clergy were commended for their harmonious relations, as such bliss appeared to be quite rare.

16 Peltonen (2006, 194–223) describes how the relations between the vicarage and parishioners could be conducted with varying degrees of animosity. The clergy could neglect their duties (e.g., schooling), or as well-connected and educated, they could offer many additional services (such as distributing medicines). Parishioners' behavior towards them could be either generous or even violent.

ples of late-eighteenth century vicars living on unrecorded credit from householders or even mixing their own funds with those of the parish granary (Suolahti 1919, 197–198, 241–247). The nineteenth century also saw a decline in societal regard for the clerical profession, most clearly seen in the sons of higher estates (including those of the clergy) choosing instead careers as civil servants or industrialists (Björklund 1939, 29–33, 133–135).

Thus, it may well be that Suolinna (1982, 93) has hit the mark when she states that parishioners (i.e., all subjects of the crown) had by the late eighteenth century learned to distinguish between practicing Christianity by its outward forms on one hand and personal piety on the other, and that this distinction was to become the rallying cry of the revivals. Ylikangas (1979, 123) interprets it as being the church orthodoxy of the time that every parishioner who went through the correct motions was a "true Christian" (e.g., administering communion was a mechanical operation with mechanistic results, regardless of the spiritual state of the recipient). During the sixteenth century Reformation the Swedish crown had taken control of the newly created Evangelical Lutheran Church and guarded orthodoxy, while the church in turn looked after state interests in the everyday affairs of its subjects. But by the eighteenth century church theology had become influenced by neology, a school of thought reflecting Enlightenment ideas that did not see a conflict between faith and reason. Christianity was advocated as a universal moral ideal with practical value, while the dogmatism of Lutheran orthodoxy was criticized (Kääriäinen, Niemelä, and Ketola 2003, 36; Hartikainen 2005, 11).

The turn of the eighteenth century, however, also saw the spread of an originally German movement dubbed Pietism (by its opponents) that aimed at church reform from the inside, to stop the spread of secularism based on ideas about natural law and philosophy. State authorities came to see this as a threat to the state-church union, and in 1726 promulgated the conventicle placard (*konventikelplakatet*; *konventikkeliplakaatti*), a statute that forbade conventicles, religious meetings in private homes that were a primary form of activity for the movement (Pentikäinen 1978, 188–190; Ylikangas 1979, 11–16; see also Ylikangas 2002, 9–10). This placard was not the first of its kind, and the authorities took care not to produce martyrs with persecution based on doctrinal issues. Instead, it was forbidden to gather in private rooms for devotions, prayer or divine service, or for preaching or interpretation of liturgic texts, with threat of significant fines, prison, or exile for organizers. Mere participation carried a threat of lesser fines. Solitary and church services were deemed sufficient for the crown's subjects. The placard also made provisions to allow churchmen to visit their parishioners' homes for purposes of inspection, but forbade them from compiling novel doctrinal treatises for this purpose without prior official approval. This last restriction was required to prevent churchmen from becoming prime movers behind these conventicles, as had already happened.

These convoluted formulations were due to the fact that there were few doctrinal differences between the state orthodoxy and the revivals aiming at individual renewal of faith. But wide interpretation of the law and onerous punishments were intended to create a deterrent effect, and only if necessary, to crush or intern the leaders. In practice deterrence was not achieved, and amendments were needed to stop enthusiastic local officials from opening legal proceedings that could turn into an unwanted spectacle (Ylikangas 1979, 17–18). Consequently, in several cases the processes were terminated by the procurator, or decisions were overturned for want of proper order in pressing the charges. Apparently in some cases conventicle organizers were punished ostensibly for some other offence, such as breaking the Sabbath or defaming the clergy. But although the placard was not struck down until the new church law of 1869, there were only a few conventicle cases after 1845. Court action had not been effective in deterring the revivals, especially the Ostrobothnian *Skörts*, and at the request of the 1842 convocation legal proceedings were to be avoided in favor of ecclesiastic reprimands (Pentikäinen 1978, 188, 192–195, 200).

5.3 Three Revivals

As described in chapter 1.3, my analysis focuses on three local revivals, those for which sufficient data regarding their practices and social composition exist. I will next introduce these revivals by reviewing what is known of their origins, and will also briefly introduce their beliefs and practices (for more on these, see section 6.2).

I address here the question concerning their notions of God and his punitiveness. These should be compared against the state church view of the time, which is commonly interpreted to have been influenced by Enlightenment neology: God was only will and reason, without heart or conscience; the devil and the afterlife were punishments rejected or downplayed; and the essence of Christianity was grounded in rationalist moral philosophy (Salomies 1962, 217–222, 269–272; Kääriäinen, Niemelä, and Ketola 2003, 36; Pyysiäinen 2004a, 182–183; Hartikainen 2005, 11). Although Salomies (1962, 271) claims that neological "extremism" was rare in Finland at the time, it was common enough in Sweden for Lars Levi Laestadius to react against it in his writings (Pyysiäinen 2004a, 182–183). Recall also the above interpretation by Ylikangas (1979, 121–123) that (in the eyes of the church) one was Christian as long as one refrained from the coarsest crimes and went through the motions of communion and other rituals. If not a *Deus otiosus*, the God of the state church certainly appeared less concerned with human affairs and morality.

5.3.1 The Jumping Revival

In the newspaper article *"Merkilliset käräjät"* ("Strange assizes," *Uusi Suometar* 234–235, October 9–10, 1895), an anonymous author recalls his childhood images of people going to church in 1833:

> Of the womenfolk, almost every other one wore dark clothing with a black headdress. Their faces were pale and their outlook gloomy, hands crossed on a hymn book that they carried against their chests. Of the menfolk, many had quite long beards, pale faces, eyes in deep thought and burning with inner zeal, and hands crossed upon their chests. [...] If this didn't come from a disease, it could only be the result of prolonged spiritual battles and temptations that had so thoroughly drained the fluids of physical life.

He goes on to describe how the church is difficult to enter with all the men and women lying on their knees and face across the floor (instead of sitting on the benches), praying aloud. He also thinks they purposefully tried to block the minister's entrance to the altar and the pulpit, and during the sermon, "deep sighs or actually blowing is constantly heard from every corner so one cannot hear the minister preach. He must interrupt his sermon to admonish the hecklers." But after a moment they continue, clearly trying to snub the minister on purpose.

> The main tenets in the teaching of the Pope of Auvainen are as follows: the inner light of the spirit is mightier than the word of God; the vessel for receiving this light is prayer, which must always be made in the most humble position. Prayers are immediately followed by hymns of praise and jumping or dancing while in a state of spiritual joy. This jumping is of a kind where the revived after their prayers suddenly jumped upwards, singing praise and raising their hands or clapping them together with such enthusiasm that they fell on the floor with exhaustion. [...] The sect's conventicles took place every Sunday evening and lasted until Monday daytime, as has been described above. All night it was dancing, hollering, and chanting, so that finally they entered rapture and, exhausted, swooned on the floor.

The author of the above text could hardly be merely reminiscing about events 62 years past, as he makes several references to other authors and official proceedings, and the image he paints of the "Jumping Revival" (*hyppyherätys*) corresponds to other depictions. This revival is thought to have started in 1817 in Eura with the blacksmith Juhana Antinpoika Uusikartano (named after his home village "the Pope of Auvainen," *Auvaisten paavi*), but didn't catch on vigorously until 1827, when it spread to several surrounding parishes, including the towns of Pori and Rauma, and when new leaders sprang up, among them shoemaker Johan Dahlberg and Michael Rostedt, another blacksmith (Heino 1976, 75–88).[17]

Leaders and charismatic members were believed to communicate God's will (Heino 1976, 113, 119–120). Uusikartano claimed to be able to tell whether an individual was destined to hell or heaven by looking into his eyes. A central concern for the Jumpers was to obtain certainty of their own salvation. Impatient emotion and seeking testimonies of the spirit were dominant; instead of pondering on the process of salvation, people wanted to know what they should do to achieve certainty of it. The answer was to seek *unio mystica* via prayer, asceticism, and ecstatic experiences, which led to some accusations of antinomianism. Akiander (1859, 274) mentions them teaching that "it is not dangerous to sin, if one later repents," but according to Heino (1976, 144–146) this belief was not taken comprehensively. After 1830 ecstatic displays were being replaced by prayer, which was performed aloud (in the manner of communal prayers) and prostrate, and though it was performed often, it happened mostly at established times and places. Regarding the prayer position, Heino (1976, 124, 127, 178) interprets it to have been necessary to humble oneself thus in order to combat the devil and the unguarded nature of humanity, so that the prayer-connection to God would not be broken. Instead of fearing constant divine surveillance, the Jumpers actively sought it.

---

17 An established view is and has been that the contemporary Supplicationism of Southwestern Finland has its origin in the folk revival begun by Liisa Eerikintytär in 1756 (Haavio 1965, 29; Salomäki 2010, 37). Contrary to this view, Heino (1976, 13, 72–73, 103, 228) argued that it wasn't until the Jumping Revival that a geographically and socially united Supplicationist movement with distinct doctrinal views was created. A connection between the Jumpers and the earlier revival is more difficult to demonstrate, due to temporal discontinuity and differences in both teaching and geographical distribution. Also, not all of the remaining "folk revivalists" embraced the new movement. Sulkunen (1999, 88–90) defends the more conventional view, emphasizing the overall unity of the revivalist phenomena in Western Finland. She sees both waves of revivalism as collective acts of defense against societal change instead of "deliberate theological stands," with the ecstatic practices of Jumpers serving to unify the earlier folk revival against challenges from invading new movements. While their dispute is ostensibly the familiar history of ideas versus structural explanation, it should be noted that Heino (1976, 22) does not deny the earlier folk revival's influence on later Supplicationism, but argues from the discontinuities that the Jumpers constituted a new (folk) revival, in conjuction with other contemporary ecstatic activities around the country – an idea that would fit quite well with Sulkunen's framework of independent revivals spreading spontaneously through the country. Also, while Heino (1976, 52–59) does not offer a cause for the birth of the Jumping Revival, he denies, for example, the influence of Herrnhutism (a form of German Pietism) in this.

Heino (1976, 165–174, 183–185, 203, 220, 223) describes the Jumping Revival stagnating after 1836 for several reasons: there was a power struggle, doctrinal disagreement, and deaths among leaders; there were court decisions that made organizing conventicles more difficult and shaped public opinion against the more outlandish practices; the proclaimed apocalypse did not arrive; and unlike in some other revivals, almost no members of the cloth associated with the Jumpers. After this, the movement solidified its ranks against the Evangelical and *Skört* movements by siding with Renqvistianism and incorporating ideas from the writings of Henrik Renqvist. Still, it lost some ground to Evangelicalism and failed to penetrate towns and cities. It seems what is now known as Supplicationism took shape fairly quickly after the end of the Jumping Revival, with the spread of the notion that full certainty of salvation may not be available in the present world, only contention for it. Ecstatic practices disappeared, but prostrate prayer, the dresscode, and long beards remained. Their denunciation of the church eased, although this was part of the general process whereby the revivals overwhelmed the church from the inside.

5.3.2 The Kuortane Revival

In 1798 Jakob Wallenberg, an unsuccessful glazier, started his own revival at Kauhava, Ostrobothnia. The following is an extract from the proceedings of a parish meeting of May 13, 1798, written by vicar Johan Gummerus (in Akiander 1860, 29–34):

> [...] juror Nils Stor-Sompi described how glazier Wallenberg has now for some months held illegal meetings, in which he instructed his followers to stomp their feet and trample the head of the snake, to which farmhand Isak and womenfolk from Sompi added that while stomping and trampling the master sang his poems, practiced foot washing during his other ceremonies, and noticed that Matts Nygård threw the Psalm book under the table. Widow Walborg Eliaedotter Kalm said she had not seen this company drunk, but she knew that they sought spirits and proclaimed as with one mouth, "Everlasting covenant!" and that Wallenberg during their services among other things ordered Matts Johansson, farmhand of Lill-Passi, to handle a woman's secret and the woman his, as those who will not do this were still tainted and had, as the order went, a "sense of shame" [*häpyrenseli*], which was to be got rid of if one was to become one of the blessed.

According to the proceedings, in addition to marching to music, Wallenberg exhorted his followers to engage in all kinds of norm breaking, while taking a lover ("heavenly bride") for himself. He had also refused to obey the orders of the local vicar, as he and his followers would not bow to temporal authorities. A contemporary newspaper article (*Åbo Tidning* 1803 issues 8–9, 13, 18, in Akiander 1860,

16–29) ascribes his teachings to Jakob Böhme's book *Psychologia vera* and describes also how his followers butchered their livestock for the group to eat, as Wallenberg had promised that he would "of the slain stock's bones bring forth living cows, back in their stalls, and turn grey stones into pieces of pure gold." These meetings attracted 20–30 persons, and only months passed before the group was apparently dying out – at least partly disappointed by the prophet's failure to make gold as promised – even before Wallenberg was taken to court and imprisoned for life (Troberg 1995, 3–4).

But as early as 1799, the movement had taken hold at Kuortane, a nearby parish. The June 9, 1799 parish meeting minutes (in Akiander 1860, 39–40) ascribe this to the "daydreamer and fantastist glazier Jakob Wallenberg's follower householder Jakob Tuomola from Lapua parish"[18] and admonish against recent "illegal meetings under the pretense of devotions with dances and games, on evenings of Sundays and holy days." But according to the accused, "their dance and games were not playful but spiritual, and they, so taken by heavenly joy, could not restrain themselves." Still, several conventicles were held in the following years, and the sectarians were reported to have threatened and ridiculed outsiders interested in their doings and disturbed church services; but after 1802, more serious threats of legal persecution forced the movement to keep a low profile as a tacitly approved part of the community (Troberg 1995, 7–13). From 1810 to 1854 the movement was led by Isak Vasu (Vasumäki), whose name was applied to the movement (*vasulaisuus*; Seppo 1987, 68).

In 1825 there was a dean's inspection for which a list of sectarians was compiled, again accused of sexual licentiousness and excessive drinking behind locked doors (proceedings of March 11, 1825, in Akiander 1860, 41–46), and in 1836 Vasu was called before the dean to describe his teaching (proceedings of September 1–3, 1836, excerpt in Akiander 1860, 46–48). By the time of this last instance the movement had become less separatistic, and the local vicar even defended it (see Seppo 1987, 71–72). Finally, reporting in the mid-nineteenth century, Akiander (1860, 38) states that still a fourth of the elderly parish population belonged to the movement. It died out completely near the end of the century (Troberg 1995, 13).

Little is known of their beliefs, but Akiander (1860, 35–37) ascribes to them the following more or less deviant tenets (possibly held at varying times, hence the contradictions):

---

18 Some other sources maintain it was brought by lodger Matts Jöransson Nygård, another of Wallenberg's followers and the first leader of the Kuortane sect (Akiander 1860, 34, 49, 53, 57), but the connection between the Kauhava glazier's goings-on and what happened at Kuortane is maintained consistently.

- Not all that is contained in the Bible is from God, but from man, though by the Holy Spirit those in the sect can distinguish between these parts (they read the Psalms, the Acts of the Apostles, and also books by the German mystic Jakob Böhme).
- Faith is but holding true the triumph of Christ. Instead of Jesus, speak only of the Lord and the Heavenly Father. Going to communion is like rending the Lord's flesh anew.
- There is no need for mending of ways; one can still mend one's ways while in-between this life and the Judgement day. For the elect, the present life is a foretaste of the heavenly joy.
- Things forbidden by God's law are not in themselves unclean or punishable. Marriage is not set by God, and people should instead live in "spiritual marriage" with their spiritual brothers and sisters.

The image of God these tenets paint is not very demanding; even things he forbids are not punishable as such, and one can redeem him/herself later in some purgatory-like state. With the lack of sources, it is more difficult to say how these regulations were interpreted in everyday beliefs of divine monitoring. The deacon of Lapua, A. O. Törnudd, later noted that for Vasu's followers, "none of the forbidden deeds in God's law were held to be really unclean or punishable, but moreso transgressions against the sect's injunctions" (Akiander 1860, 50); however, this does not tell us whether these injuctions were enforced with threats of divine retribution.

5.3.3 Ostrobothnian *Skörts*

In a February 10, 1839 letter to his colleague, county constable (*nimismies*) V. Sundström described the revival that was taking place around Nivala (Rosendal 1905, 89–90, n. 2):[19] 1) The sectarians distinguished themselves from other folk by their clothing: coats with long tails, plain sweaters, women using black headscarves instead of hats, and men wearing brimless caps. 2) Their devotions consisted of reading spiritual literature (especially *"en ropandes röst i öknen,"* "A Voice in the Wilderness," apparently the 1823 book by Peter Jonsson Topp) and singing spiritual songs (from the book *"Sijons sånger,"* "Songs of Zion"), and this they took to be an important service that was not to be interrupted. 3)

19 The Swedish of this county constable is quite broken, so direct translation is not attempted here.

Every Saturday afternoon people from all parts of the parish came to the vicarage, and those from more distant parts lived by the charity of others. 4) They also held feasts or devotions in various places during workdays, which took a considerable amount of time. 5) Temporary preacher (*Interims Predikanten*) N. G. Malmberg installed in the doorway of the vicarage a collection box for missionary work, into which womenfolk donated silk clothing and gold rings among other things, which were then sold. While the above letter was apparently a solicited one for the purposes of court action, the revived, in their defense, did not deny any of the basic facts stated above but rather disputed the intimations that this had led to factionalism, sloth, or misappropriation of the donated funds; instead they presented a view that people who had previously spent their time and money on gambling, alcohol, and fancy clothing had now mended their ways and suffered from a spiritual thirst the additional devotions were intended to quench (Rosendal 1905, 114–119, 123–126).

This was not an isolated occurrence, but similiar reports were coming also from other parts of Ostrobothnia (Rosendal 1905, *passim.*). A traditional view has it that the Ostrobothnian *Skört* Revival was instigated by local clergy following their personal awakenings, namely Jonas Lagus at Ylivieska and Nils Gustav Malmberg at Kalajoki from 1828–1831 (Rosendal 1902, 255–281), and later in the decade by Fredrik Östring at Uusikaarlepyy, Frans Oskar Durchman at Ylihärmä and Isokyrö, and Malmberg also at Lapua (Rosendal 1905, 143–159). Ylikangas (1979, 29–41 and *passim*; 1990a, 322; 2002, 18) has forcefully argued against this view. When testifying in court as defendants, these clergymen claimed not to be the instigators of these revivals, but only to have offered guidance to parishioners whose desire to hear the word of God had spontaneously increased (Ylikangas 1979, 47, 104). That this is not merely courtroom tactics is supported by letters they sent to each other. Lagus describes to Jakob Wegelius in 1832 how he had tried to guide the revived to the right track, discouraging ecstaticism and conventicles outside the rectory. A second letter, dated 1836, states that "[t]he teachers were rarely prepared for the revival and spiritual movement that took hold of the people." In this view, the revived clergymen were different from others only in being up to the task laid upon them by the laity,[20] but no doubt they also gave the movement a drive and authority it would have other-

20 Reviewing the court proceedings begun against those involved, Ylikangas (1979, 196) has come to the conclusion that church leaders wanted (with curious bending of the law on several occasions) to spare the accused clergymen from ruination in order to keep the revival in their hands. This would have to mean that they saw the revival by this time to go on with or without clergy.

wise lacked (Ylikangas 1979, 198–199, 207–208; similiarly in Ruutu 1967, 102–103).

The early history of Ostrobothnian *Skört* Revival is entangled with that of the Savonian revivals. The Northern Savonian movement that had begun in 1796 soon came under the lead of Paavo Ruotsalainen, a householder (Rosendal 1902, 3–33). His travels took him also to Ostrobothnia, where he met the local revived, and eventually (circa 1835–1836) was recognized also by the clerical leaders there as the spiritual authority (Rosendal 1902, 290–299; Ylikangas 1979, 234–239). But Ylikangas (1979, 201–203) interprets that to mean it was the Savonian influence that had already "re-ignited" what heritage remained of the eighteenth century mystics and separatists (chapter 5.1; Luukkanen 2001) into a new revival. A clear example of this influence would be the spread of the Savonian *skört* in the 1820s, an austere and old-fashioned peasant dress that gave the movement both a name (*Skörts*; *körtit*) and a much-maligned public image. When Elias Lönnrot, the district physician in Kajaani, published a diatribe against the ways of the local revived in 1835, even those of the Ostrobothnian clergy who had initially opposed the distinctive clothing undertook a defense of it, indicative of similiarity and affinity between Ostrobothnia and Savonia (Ylikangas 1979, 219–225, 237–238).

Despite the furor over these and other outward practices it seems there was little in the beliefs of the *Skörts* per se that was deviant from state church orthodoxy (Ruuth 1942, 142). It was rather the new forms of social activity and demands of personal awakening that were disturbing to the established ways. However, Paavo Ruotsalainen also went to great lengths to emphasize that human beings could not trust in their own deeds in obtaining salvation but had to put faith solely in the greatness and mercy of God (Niskanen [1837] 1998, 43; Haavio 1965, 40; Remes 1995, 37–40, 79, 222–223), as did the *Skört* ministers (Ylikangas 1979, 137). The need to stave off hypocrisy and self-righteousness gave some licence that manifested occasionally in uninhibited drinking (Rosendal 1912, 70; Hirvonen 2003, 31–32), but this did not mean that sin was taken to be nonexistent. As will be described in more detail in section 6.2.2, a common theme with *Skört* ministers was to emphasize the sinfulness of those present, even that of the speaker himself, and also actively threaten their parishioners with divine punishment and hellfire. For example, the Mustasaari deacon Kustaa Vilhelm Lybeck once finished his sermon with the words, "You will all end up in hell, amen" (Rosendal 1905, 463). In court, Vöyri deacon Johan Mikael Stenbäck was accused of pressuring a dying woman into confessing theft and adultery, and then denying her communion until she had repented sufficiently (Ylikangas 1979, 120–121, 136–138). Even if human beings could not buy salvation with good deeds from the God of the *Skörts*, a good Christian had to live a careful life in order to obtain God's mercy.

The numerous court processes instigated against the *Skörts* (see chapter 6.1.3) did not bring their revivalism to a halt. More threatening were the internal divisions that arose mid-century. First to split in 1844 were the followers of clergyman Fredrik Gabriel Hedberg, who went on to form what is known as the Evangelical movement (*Evankelinen herätysliike*; Haavio 1965, 47–48, 55–56). After the death of Paavo Ruotsalainen in 1852, leadership issues created more divisions, and several clergymen detatched themselves from the *Skört* Revival (Ylikangas 1984, chapter 1). A time of lay leaders and relative stagnation followed, but after the turn of the century the *Skört* movement gained in importance as it became associated in the public eye with nationalistic ideals (Huhta 2007, 224–227). In the twenty-first century it has lost its strong cohesiveness, becoming more like a loose network centered on the annual summer festival, yet it still exerts a strong influence inside the state church (Salomäki 2010, 43, 375, 378).

6 From the Deprived to the Revived: A Self-Organizatory Process

> "Whenever messengers are sent, they have prestige among their people." That means that such a man has group feeling and influence which protect him from harm at the hands of unbelievers, until he has delivered the messages of his Lord and achieved the degree of complete perfection with respect to his religion and religious organization that God intended for him (Ibn Khaldûn 1967, 71).

6.1 Social Composition of the Revived

In the first decades of the nineteenth century there were a number of mass court cases taken against revivals on the basis of the anti-conventicle legislation. These cases allow us to catch a more systematic glimpse of who the revived were. There were two sessions against the Jumpers from 1831–1833, a number of sessions against the Ostrobothnian *Skörts* in the 1840s, and those that partook in the Kuortane Revival were rounded up twice for inspection (from 1799–1802 and again in 1825).

In reviewing these cases I am relying on original research by Heino (1976), Ylikangas (1979), and Troberg (1995), who have compiled the material from various archives. My contribution here is twofold: First, to my knowledge no one has yet put together the material on these three revivals and performed a comprehensive analysis of them. Second, the analyses by the aforementioned historians lack tests for statistical significance. In other words, none of them have controlled for the possibility that the social composition of the revived did not actually differ from that of the general populace, with random variation being the reason for skewed distributions (type I error). While they generally base their reasoning on rather glaring differences, the oft-mentioned gender imbalance becomes suspect when properly tested against the actual population distribution. I am thus supplementing existing research by conducting these tests before proceeding to an overall analysis. In order to do this, I was at some points forced to re-compile the figures either from the appendical information in Heino (1976) and Troberg (1995) or from original sources; Ylikangas (1979) is the most explicit and thorough in his use of the archival material, so I have used his work as is. In each case I have tested the goodness of fit between the revived and general population with categorization based on social class and gender, both separately and in combination. Also, I have tested whether there is any significant interaction between social class and gender among the revived samples (e.g., whether women but not men of some particular class were more drawn to the revival than other women), and some other tests when necessary.

This kind of significance testing for categorical distributions from which no normality assumptions can be made is usually performed with Pearson's χ^2 goodness-of-fit test, which produces a likelihood on drawing a given observed distribution from some expected (theoretical) distribution (Howell 2002, 146–148, 158–159), the null hypothesis being that the observations constitute a random sample of the population. But for this test to be applicable, the expected frequencies for each category should be at least five (a conservative position), or between one and five in no more than 20 percent of categories (a less stringent position); the data at hand seriously conflict with these requirements in many cases. One solution is to use Monte Carlo simulations (a suggestion from Kimmo Vehkalahti, personal communication on October 15 & 19, 2010) that work in this case by repeatedly creating a random sample from the given theoretical distribution and calculating the χ^2 statistic for this pseudo-observation; here the p-value is the proportion of the obtained statistics that are greater than or equal to the analytical result (which is included in the tally). Below I report the analytical p-value only when the requirements for calculating it are met, and otherwise the simulated value obtained with a Monte Carlo test based on 10,000 samples. Degrees of freedom are applicable only in the analytical case, so I will omit them from simulation results to tell the two apart. Tests of independence I have conducted with Fisher's exact test on the null hypothesis that there is no interaction between gender and social class in the samples. All calculations have been performed using the R statistical package (version 2.12.1 for Debian GNU/Linux).

6.1.1 The Jumping Revival

The Jumping Revival led to two court sessions against 106 Jumpers mainly from Luvia and 27 from Kodiksam village in Lappi parish. Apparently the former consisted of all adult inhabitants of the chapel area known to have participated in conventicles (autumn assizes August 23, 1831), while the latter is related to a surprise raid on a single household conventicle (July 29, 1832). Heino (1976, 90–92, 96–97, 104–105) reviews the social composition of these groups, but since some of his figures are not amenable to a more detailed analysis, I have been forced to recompile them.[1]

1 Heino (1976, 261–265) provides lists of the accused as an appendix, and distribution of the

Table 1: Social composition of Luvian Jumpers (compiled from Heino 1976, 261–264 and VÄ113).

	revived				population			
social class	men	women	total	%	men	women	total	%
gentry, clergy & civil servants	0	0	0	0.0	5	13	18	1.8
householders	5	8	13	12.3	56	56	112	10.9
tenant farmers	4	3	7	6.6	11	11	22	2.1
crofters	16	14	30	28.3	140	140	280	27.3
hirelings	5	10	15	14.2	136	162	298	29.0
craftsmen	5	6	11	10.4	16	12	28	2.7
lodgers	9	14	23	21.7	72	64	136	13.3
soldiers	0	0	0	0.0	5	5	10	1.0
apprentices	0		0	0.0	34		34	3.3
widows		7	7	6.6		79	79	7.7
the poor	0	0	0	0.0	3	6	9	0.9
total	44	62	106		478	548	1026	

Goodness of fit between distributions based on:
- gender and social class: $\chi^2 = 54.0648$, $p = .0002$
- social class: $\chi^2 = 61.0826$, $p = .0003$
- gender: $\chi^2(1) = 1.099$, $p = .2945$

Dependency of gender and social class among revived: $p = .2157$

Table 2: Social composition of Kodiksami Jumpers (compiled from Heino 1976, 265 and the Lappi communion book 1832–1838).

	revived				population			
social class	men	women	total	%	men	women	total	%
gentry, clergy, & civil servants	0	0	0	0.0	1	1	2	1.8
householders	3	5	8	47.1	22	20	42	37.2
crofters	2	1	3	17.6	11	14	25	22.1
hirelings	1	3	4	23.5	14	13	27	23.9
craftsmen	0	1	1	5.9	4	1	5	4.4
lodgers	1	0	1	5.9	2	3	5	4.4
widows		0	0	0.0		7	7	6.2
total	7	10	17		54	59	113	

Goodness of fit between distributions based on:
- gender and social class: $\chi^2 = 11.9672$, $p = .4016$
- social class: $\chi^2 = 2.1196$, $p = .9231$
- gender: $\chi^2(1) = 0.2978$, $p = .5853$

Dependency of gender and social class among revived: $p = .6501$

The mean age of Jumpers from two Luvian villages (Korpi and Niemi) in 1830 was 37.4 years ($\sigma = 7.2$), while for the general population of over-15-year-old Luvians it is 38.8 ($\sigma = 17.2$).[2] There is thus no difference in means, but the Jumpers are clustered around the mean more strongly. As Heino (1976, 105) states, every third 30 – 45 year-old Luvian was being charged for breaking against the placard. In 1832 Kodiksami the Jumpers' mean age is 28.7 years ($\sigma = 10.3$), while for the general population (excluding minors) it is 34.2 ($\sigma = 15.8$).

Looking at the social profiles in tables 1 and 2, in both cases there is a small majority of women, but the difference is not statistically significant. Neither is there significant dependency between gender and social class. As to the class distribution, in Luvia the noteworthy differences are the over-representation of craftsmen and lodgers, and the under-representation of hirelings. The overall difference is statistically significant. However, at Kodiksami there are no substantial or significant differences from the general population. Heino (1976, 109) notes that in addition to being strongly represented, the craftsmen occupied prominent positions in the movement as leaders or conventicle organizers. The latter applies also to the masters of households, of whom all were accused of organizing conventicles.

Luvian general populace in 1830 is presented in the population tables (microfilm VÄ113 at the National Archives of Finland, spreads 107–109). Since the underaged were clearly not prosecuted at Luvia, I have excluded them also from the population figures. But the table includes two problematic lumpings: 1) §e includes all kinds of agricultural workers, "old and feeble householder, crofter and lodger menfolk who are maintained with agriculture" (48) to which there are no corresponding designations among the accused. I decided to divide this figure in proportion among householders, tenant farmers, crofters, hirelings, and lodgers. 2) The "wives [...] of men who belong in §e" (262) cannot be separated into subcategories as precisely as men are. Here I first subtracted numbers equal to men who were householders, tenant farmers, or crofters (since these classes rarely went unmarried; Haatanen 1968, 42). The remainder was distributed in proportion to categories of lodgers and hirelings, in addition to figures for unmarried women in these classes.

As for the Kodiksam case, the tables for Lappi are missing from the archives (personal communication with Pirkko Vartiala, Provincial Archives of Turku, November 1, 2010). Here I have reconstructed the Kodiksam population from the 1832–1838 communion book for Lappi, digitally archived by Finland's Family History Association. Of the accused at Kodiksam, I have discounted the five inhabitants of the house other than the master, as their actual participation is uncertain, three persons who can be identified from the books of neighboring villages (cavalry farm master Johan Isac Mattson Suckala, lodger Hedvig Johansdottir Orre, and 11-year-old farmer's daughter Ulrica Riikilä, whose presence makes one wonder), and two persons who cannot be identified as Kodiksami residents (lodger's wife Maria Eriksdottir and maidservant Maria).

2 The population values were calculated from the 1830 population table, which includes an age-group breakdown in categories spanning five years.

6.1.2 The Kuortane Revival

The data we have on the Kuortane Revival form an interesting case due to a generational gap between the two court cases (98 accused from 1799–1802 and 87 in 1825). Thus we can glimpse how their social composition changed in the course of the revival. Social profiles of both the revived and the general populace at these two time periods are presented in tables 3 and 4 with results from significance tests.[3]

In 1799 the mean age of the revived was 33.8 (σ = 14.8), and that of the 1805 general population was 36.8 (σ = 16.0), excluding those under 15; in 1825 it was 44.0 (σ = 13.7), compared with 37.2 (σ = 15.9) for the general populace. The earlier difference is not significant (χ^2 = 15.828, p = .2511), while the later one is (χ^2 = 37.0875, p = .0127). This change cannot be attributed to natural aging of the partakers as the number of individuals who stood accused at both proceedings is only thirteen, a surprisingly low number.[4] Even with a generation-wide time

---

3 The quinquennial population table of occupational distribution for the year 1800 is missing from the archives. The Finnish National Archives microfilm VÄ177 has tables from 1802–1847, and those at the digital archive of Finland's Family History Association from 1774–1798, with 1785 being the last year from which there are occupational distributions (spreads 49–50). The microfilm VÄ178 stated as the source in the digital archive is apparently not the same as the one in the National Archives, as the latter contained material from a later period only. For these reasons I have used the population data from year 1805 in table 3.

The population table for 1825 is present on microfilm VÄ177, but Troberg (1995) has used the wrong one, probably due to the film containing tables from the mother parish (*moderkyrka församling*) and the greater parish (*pastorats församlingar* that includes Alavus and Töysä) both under main heading "Kuortane," but in a very confused order. Otherwise his table 9 has figures from the mother parish, but the figures for 1825 (that show 200 %–500 % raises from 1820) are from the greater parish area. That this is an error is corroborated by his 1825 population total (Troberg's table 2) being incongruent with this sudden raise. Thus, I have used the proper figures from VÄ177 in table 4 (below).

A more difficult problem is that the court proceedings, unlike the census, do not differentiate between the elderly and the others. For purposes of comparison, I have recoded some of the accused as "elderly" if they fulfilled either criterion 1 (age of sixty or older and social status such as "widower" or "old master") or criterion 2 (age of seventy or older). From 1799–1802 this meant individuals 7a and 62, and in 1825 individuals 17, 57, 58, 64, 89 and 119 (numbers refer to the appendix in Troberg 1995), which leaves the proportion of elderly among the revived very low from 1799–1802 and could create a statistical artifact. However, this is in keeping with their generally young age profile at that time, and from 1825–1826 the proportion is much closer to that of the general populace.

4 Troberg (1995, 35) states eleven, which is odd, since in his own appendix individuals 2, 35, 47, 55, 58, 59, 61, 64, 75, 104, 119, 140 and 159 are clearly mentioned as standing accused during both prosecutions.

Table 3: Social composition of the Kuortane Revival from 1799–1802 (compiled from Troberg 1995, appendix 1 and VÄ177, spreads 9–10).

	revived				population 1805			
social class	men	women	total	%	men	women	total	%
hirelings	3	11	14	14.3	242	271	513	45.0
crofters	5	12	17	17.3	60	60	120	10.5
lodgers	0	6	6	6.1	28	22	50	4.4
soldiers	1	3	4	4.1	20	17	37	3.2
householders	15	40	55	56.1	113	113	226	19.8
craftsmen	0	0	0	0.0	3	0	3	0.3
gentry, clergy, & civil servants	0	0	0	0.0	9	7	16	1.4
elderly	1	1	2	2.0	59	76	135	11.9
the poor	0	0	0	0.0	22	17	39	3.4
total	25	73	98		556	583	1139	

Goodness of fit between distributions based on:
- gender and social class: $\chi^2 = 154.1685$, $p = .0001$
- social class: $\chi^2 = 103.7357$, $p = .0001$
- gender: $\chi^2(1) = 21.3016$, $p = .000003924$

Dependency of gender and social class among revived: $p = .6683$

Table 4: Social composition of the Kuortane Revival from 1825–1826 (compiled from Troberg 1995, appendix 1 and VÄ177, spreads 97–98).

	revived				population 1825			
social class	men	women	total	%	men	women	total	%
hirelings	1	2	3	3.4	228	228	456	43.0
crofters	6	11	17	19.5	78	78	156	14.7
lodgers	1	1	2	2.3	28	21	49	4.6
soldiers	0	0	0	0.0	10	4	14	1.3
householders	26	32	58	66.7	117	117	234	22.1
craftsmen	0	0	0	0.0	1	0	1	0.1
gentry, clergy, & civil servants	0	1	1	1.1	7	6	13	1.2
elderly	3	3	6	6.9	26	80	106	10.0
the poor	0	0	0	0.0	6	26	32	3.0
total	37	50	87		501	560	1061	

Goodness of fit between distributions based on:
- gender and social class: $\chi^2 = 123.6551$, $p =.0001$
- social class: $\chi^2 = 117.2309$, $p = .0001$
- gender: $\chi^2(1) = 0.7681$, $p = .3808$

Dependency of gender and social class among revived: $p = .9692$

gap the movement had not apparently turned from revivalism into foster-religiosity. More interestingly, its pull was not determined by age, but rather by the time of birth, most specifically the 1780s (Troberg 1995, 89).

Another interesting issue is gender. As seen from tables 3 and 4, women were grossly over-represented from 1799–1802, but from 1825–1826 the difference is no longer statistically significant. Testing for the significance of this change with Fisher's exact test yields p = .01589 (after removing from the tally those 13 individuals who partook in both revivals), so we can conclude that while the revival initially had considerably more attraction for women, with time this difference disappeared.

As to their class composition, householders and crofters are clearly over-represented groups from 1799–1802, while hirelings and the elderly are clearly under-represented. Looking at the tax roll of 1800, Troberg (1995, 85–89) states that while the most affluent individuals were absent, sectarians were slightly more affluent than the average parishioner. From 1825–1826 the situation is similiar, but now householders were the single, even more strongly over-represented group, while hirelings are conversely almost absent. Note that gentry, clergy, and civil servants are absent, which is statistically not too surprising, but should be kept in mind.[5] At both times, differences between distributions are statistically significant, and also change in class composition over time appears significant (for this Fisher's exact test gives p = .01597). But at neither observation point are gender and social class significantly associated: the pull of the revival was determined by social class and gender working separately, not in interaction.

6.1.3 Ostrobothnian *Skörts*

The court processes were most numerous in late 1830s Ostrobothnia. However, their representativeness varies, so I will only briefly comment on most of them.[6] The first of these was from 1838–1840 at Kalajoki, with five accused clergymen and 75 others; while this sample is probably not representative, as the prosecution aimed at creating an example and especially hounded those responsible for organizing the conventicles, it should be noted that the overwhelming majority (68.8 %) were householders (Ylikangas (1979, 42–50).

---

5 The single member of the clerical estate among those prosecuted in 1825 was the cantor's wife; while her place among the estates is debatable (see Wirilander 1974, 98–99), it fits with the general pattern of women with associations to the lower clergy joining the revivals (see below).

6 For an English language reference on findings reviewed in this subchapter, see Ylikangas (2002, 11–17).

Table 5: Social composition of the revived at Kirkonkylä and Forsby villages of Uusikaarlepyy parish, 1838–1839 (adapted from Ylikangas 1979, 296).

	revived				population			
social class	men	women	total	%	men	women	total	%
the elderly	2	2	4	5.6	3	2	5	1.3
householders	21	22	43	60.6	81	91	172	45.9
crofters	1	2	3	4.2	18	20	38	10.1
craftsmen	3	3	6	8.5	3	5	8	2.1
hirelings	0	2	2	2.8	23	38	61	16.3
lodgers	2	6	8	11.3	30	40	70	18.7
others	1	4	5	7.0	6	15	21	5.6
total	30	41	71		164	211	375	

Goodness of fit between distributions based on:
- gender and social class: $\chi^2 = 42.6828$, $p = .0017$
- social class: $\chi^2 = 39.1612$, $p = 0.0002$
- gender: $\chi^2(1) = 0.0632$, $p = .8015$

Dependency of gender and social class among revived: $p = .6828$

The second proceedings with more systematic summonings were at Uusikaarlepyy (Swedish *Nykarleby*) from 1838–1841 against 212 revived (a figure which may be close to their actual number in the area[7]), a process Ylikangas (1979, 55) claims to be in its size without parallel in Finnish legal history. I will thus go over these data in detail.

The accused were disproportionately from the town: 48 townsfolk made up 6.5 % of a corresponding populace of 742 (in 1838), while for 158 country dwellers from the Uusikaarlepyy parish (pop. 3686 in 1838) the percentage is 4.3 %. This difference is statistically significant ($\chi^2(1) = 6.3242$, $p = .01191$). Of the 48 townsfolk, only 16.7 % (8) were men, but of 164 country dwellers 45.7 % (75) were men, a significant interaction between gender and place of residence ($p = .0002083$). Ylikangas (1979, 64, 71) reports the average age of townsfolk to have been lower, with 35 % born in or after 1810 (of countrymen 26.2 %), but the total mean age as 38 years, which he considers to be quite high but nevertheless explainable by their generally high social status (see below).

7 Temporary preacher (*armovuodensaarnaaja*) Fredrik Östring, charged as the leader of the revival, complained in court of the impossibility of teaching 300 people individually (a palpable overestimation), and reported to have sold 200 New Testaments. Individual conventicles attracted an average of 30–40 people, or 100 at most. See Ylikangas (1979, 87).

Table 6: Social composition of the revived at Uusikaarlepyy town, 1838–1839 (adapted from Ylikangas 1979, 297).

	revived				population			
social class	men	women	total	%	men	women	total	%
gentry & civil servants	0	7	7	14.6	23	31	54	7.0
aldermen & mayors	0	0	0	0.0	10	9	19	2.4
merchants & burghers	1	1	2	4.2	11	16	27	3.5
skippers & mates	2	2	4	8.3	28	25	53	6.8
shop staff	0	0	0	0.0	32	4	36	4.6
craftsmen	4	7	11	22.9	39	45	84	10.8
sons & daughters	1	7	8	16.7	13	75	88	11.3
seamen	0	13	13	27.1	147	97	244	31.4
hirelings	0	2	2	4.2	31	87	118	15.2
others	0	1	1	2.1	21	32	53	6.8
total	8	40	48		355	421	776	

Goodness of fit between distributions based on:
gender and social class: $\chi^2 = 50.3872$, $p = .0022$
social class: $\chi^2 = 21.0411$, $p = .0165$
gender: $\chi^2(1) = 16.3556$, $p = .0000525$
Dependency of gender and social class among revived: $p = .04093$

To analyze their occupational composition against that of the general populace, Ylikangas (1979, 64–65) argues that regarding the countryside, it is best to concentrate on the adult population in those villages the movement had sufficient time to penetrate thoroughly, namely Forsby and the church village. Figures for these are given in table 5, and for the town in table 6. In both cases the differences to general population are statistically significant. In the countryside the revived were clearly over-represented among craftsmen, pensioners, householders, and "others." Almost every member of the two first-mentioned groups joined the revival. It is of note that "others" consist almost completely of lower clergy and their families (Ylikangas 1979, 67–68). Under-represented were crofters, hirelings, and lodgers. Among townsfolk, craftsmen were again over-represented, but so were the "higher" groups: here it is important to note that the latter consist mostly of women from the lower rungs of the high society.[8] The under-represented include aldermen, mayors, shop staff, seamen, and hirelings. As Ylikangas (1979, 69–70) concludes, in both town and country the Ostro-

8 The three men (a burgher, a skipper, and a mate) were each accompanied either by their wives or daughters (Ylikangas 1979, 69).

bothnian Revival drew its support from the more affluent social groups, the (upper) middle class of the time, even though those in the highest positions not only shunned it but also actively tried to ban its spread. The over-representation of women is consistent with this, as they were at the time barred from the highest societal power.

The third Ostrobothnian court case is from Vöyri, where a conventicle was raided in 1840 (Ylikangas 1979, 124–131). Of the "at least sixty" persons inside, a group of women (most likely of householder class) managed to escape before those present could be identified. Still, the remaining group of 48 persons offers a corroborative sample to that of Uusikaarlepyy. As the proportion of lower classes in the Vöyri sample is even lower than at Uusikaarlepyy, there is less reason to doubt that this had resulted from lack of enthusiasm in persecuting them. 72.9 % of those present were householders (again a gross over-representation), young masters and their wives being most prominent, while the landless are almost completely absent. The other noticeable group is that of women with ties to clergy. After the raid, people were ordered to go to their homes, but those who stayed in defiance of this were mostly householding males.

There was also a second court process at Uusikaarlepyy, the last mass court case against the revived or any other religious movement in Finland (Ylikangas 1979, 170–180, 190). It began in 1840 with a list of 71 accused, compiled apparently of those well-known to be revived, with some purposefully omitted. Again householders are disproportionately prominent (69.0 %), while the proportion of gentry has also risen (16.9 %). While not all of the latter are clergy, almost all have familial ties to them.

It is noteworthy that during the first proceedings at Uusikaarlepyy, the revived front did not hold, but some individuals tried to save their own skin by implicating others (Ylikangas 1979, 61–62, 70–73, 178). Most of the townsfolk (73 %) plead guilty, and only 4 % denied the charge completely, while in the country the numbers are 39 % and 38 %, respectively. Women were slightly more inclined to plead guilty, as were those of lower social groups (in the countryside) and younger individuals. It appears that this behavior can be understood in relation to the length of engagement with the revival: those who had longer association with it (women, lower classes, those from parish heartlands) had stronger convictions. Possibly they also had less to lose than the householder men who appear as the turncoats here. Their behavior during the second Uusikaarlepyy process was similiar.

6.1.4 Comparison of Social Profiles

Based on the findings above, several conclusions can be made:

- The revived consisted in every case mainly of individuals in their late thirties, with both youngsters and the elderly absent. Data from the Kuortane Revival revealed the determinant to be time of birth rather than age, but whether this applies to the other cases cannot be determined.
- Gender imbalance (in favor of women) is significant only in the earlier phase of the Kuortane Revival and at Uusikaarlepyy. The Kuortane data, observations by Ylikangas (1979, 70 – 73), and the anecdotal evidence concerning female preachers (see chapter 5.1) indicate that there may be a general pattern of women having been more drawn to the revivals in their beginnings, with men following in their wake. Also Rosendal (1902, 270) refers to early female dominance "almost everywhere."
- The lowest and highest echelons of society shunned the revivals: hirelings were relatively absent in every case covered. Lodgers had a strong presence only at Luvia. When gentry, clergy, burghers, or civil servants appear to have participated, they are revealed to be women or less prominent members of these estates.
- Householders were disproportionally active in the movement, and their involvement appears to have only strenghtened over time. In some cases we also have evidence that these farmers were those of their rank who either had trouble fulfilling their aspirations or were particularly enterpreneurial (Ylikangas 1979, 273, 281; Troberg 1995, 87, 96; see section 7.1.1). Also, craftsmen were over-represented except at Kuortane, where this is partly explained by their almost complete absence from the general population at the later observation point.[9]
- Crofters' interest in the revivals varies with region: they were prominent at Kuortane alongside householders, equally represented among Jumpers (as were householders), but do not show up in proportion to their numbers among the Ostrobothnian *Skörts*. It should be noted that the actual position of crofters also varied by region. Jutikkala (1958, 371) states that in southwestern Finland the crofters were wealthier and their position as tenants more secure than elsewhere, so their plight may not have been that different from that of householders. At Kuortane the line separating a householder

9 This is not exceptional in itself, as it was not uncommon for people to craft their clothes, shoes, and other utensils themselves, eliminating the need for professional craftsmen in some locations (Ranta 1978, 53).

from a crofter was thin due to several factors (see section 7.1), which made for fast social mobility between the two classes. To clarify this issue further requires a review of the tenancy contracts of crofters who joined the revivals and those who shunned it, which I leave for future research.

The cases covered in this subchapter include to my knowledge all data sources on the early phase of the revivals that can be considered reasonably systematic and representative, that is, mass court cases. This results in the exclusion of large-scale revivals in northern and eastern Finland (Savonian *Skörts*, Renqvistians, and Laestadians) from the analysis. But some anecdotal observations and historians' analyses can be cited to support the notion that the general pattern in these revivals fits with conclusions made above.

The "ecological hypothesis" that the revivals had the strongest pull among the land-owning populace was first put forth by Ruutu (1967, 96, 100–101), who observed their spread on the national map and found that they overlap with those areas that had a longer tradition of individualism and liberty, and where independent householders were strongly represented: the southwestern coast, Ostrobothnia, Savonia, Northern Karelia, and Lapland. The crofter-dominated areas of Tavastia and Satakunta remained mostly blank, as did Southern Karelia around Vyborg. The last-mentioned "Old Finland" (*Vanha Suomi*) constitutes an interesting further test on his hypothesis. The Laestadian Revival *did* spread to this area from 1875–1890 (Kinnunen 2004, 170–171; Ruutu had limited his analysis to 1850). This coincides with the Finnish government helping the local householders from 1871–1891 to redeem themselves from being virtual slaves to their Russian liege lords (Jutikkala 1958, 358–368).[10] I find this to support Ruutu's contention that it was the prevalance of independent householders in the area that was determinative in whether a strong revival took hold, but this does not seem to have required very long traditions in liberty.

10 This part of the country had been enjoined to the Russian realm already in the Great Northern War (*suuri pohjansota*) of 1721, and the "Hats' Russian War" (*hattujen sota*) of 1743, and even though in 1812 it was again made part of the now Russian Grand Duchy of Finland, the province was left with a legacy of substantial fiefs granted to the Russian nobility. Jutikkala (1958, 358–368) describes that travelers coming from other parts of the country commented on the poverty and primitiveness of agriculture, but most serious for householders in these areas was the threat that the sensibilities of their new masters imposed on their independence. These counted their dependents (in the Russian way) as "souls" or even "slaves" over whom they had much wider judicial powers than their Swedish counterparts had had, and they could evict unruly peasants from their farms. This situation was slowly remedied between 1871–1891, when the Finnish government helped the householders buy back their old property.

More recently, Kimanen (2008, 35–39, 46–49) has used an ingenious method to reconstruct the 1820s Renqvistian Revival at a single Karelian village. As with some other revivals, there was a sudden upsurge in the communion activity of those involved, and as these visits were recorded in the communion books, these can be used to identify those who openly partook in this norm-deviant behavior. However, a problem remains: there is no natural threshold at which to consider a person an overactive communion-goer. Kimanen has been perhaps overly cautious in this, and her criterion of four visits in at least two years results in a sample of 28 individuals. Their social composition does not conform to the pattern I have described above, with householders lacking and hirelings overrepresented. However, if this sample were to be expanded to include also those who took communion at least three times in three years, the additional 30 persons would consist almost completely of householders – a composition that seems to fit quite well with what has been observed above. Hartikainen (2005, 224), another Renqvistian scholar, found that of those (73) who signed petitions for Renqvist, almost all (60) were householders or sons of householders. But as he duly notes, it would have been natural to collect signatures mostly from those of higher standing, even if they were in a minority. However, as Kimanen for the most part restricts her analyses to the smaller sample, they are not useful for the purposes of the present study. Most interesting is her observation that the somewhat less conspicuous communion activity of householders may indicate that they were more cautious since they had more to lose, as it fits the observations of Ylikangas (1979, 70–72, 178) concerning the *Skörts* pleadings in the courtroom.

Concerning the Laestadian Revival that arose in northern Sweden and Finland around the mid-nineteenth century, Raittila (1976, 158–159, 236–237) has compiled information on the early lay preachers. While it would be untenable to assume them to represent the movement as a whole, the very high proportion (17.1 %) of craftsmen among them merits attention, and the householder majority (52.9 %) does not conflict with what has been presented above. During their fast expansion in the 1860s, the strongest centers of Laestadian activity were the well-connected and literate areas on the Torne River and Bothnian Bay. Kinnunen (2004, 166–168, 171, 350–357) has studied the revival as it arrived in Karelia in the late nineteenth century. Again, it found strongest support in more populous parishes, but even more so in towns. Their class composition included in the countryside mostly householders, and in towns enterpreneurial craftsmen and merchants, but also landless people. A notable feature in towns was that many of them had migrated large distances from their birthplace, despite which some rose rapidly into prominent positions in a trade (no wonder, as "the Laestadians have been very active in promoting the interests of craftsmen and small industries," 356). Also the study by Talonen (1993, 15, 42, 55, 59–62, 67, 128–129, 151) of a late Laestadian branch known for its ascetic rigor (*esikoislestadiolaisuus*) in the early twentieth century emphasizes their enterpreneurial activity and ability to provide work and make connections for co-religionists who migrated from the north.

The Evangelical movement that split from the *Skörts* in 1843 was later dominated by merchants and city gentry, especially clergy (Koskenniemi 1967, 62, 110). But these data are not quite comparable to early observations we have on the other revivals, as they come from the registers of the Evangelical Society, founded only in 1873 as a closed society.

Comparisons can also be made to contemporary revivals in other Nordic countries. Studying the court actions against *läseriet* ("Readerism") of Swedish Norrland from 1800–1850, Ericsson and Harnesk (1994, chapter 8) formed a corroborative picture of their social composition. It too was dominated by well-to-do householders (over time even more so), with the *obesuttna*, landless dependent groups keeping their distance. Ylikangas (2002, 24–26, 28) sees a strong correspondence also between Ostrobothnian *Skörts* and the Haugean movement in Norway. Although initially a peasant movement (it did not attract the lower social groups), it spread also to cities and by the 1830s had become a nationwide phenomenon. Haugeans also demanded more freedom in ecclesiastical and religious matters, but the connection to entrepreneurial ethos was even clearer. The Haugeans engaged themselves (for the most part successfully) in trade, seafaring, and setting up various kinds of mills and factories, emigrating when necessary. "Strikingly often, a member of Hauge's brotherhood was a shopkeeper or an artisan who had moved to his present abode from elsewhere with the intention of spreading the faith. He thus often lacked a fixed social position or acceptance, although through his industriousneess, thrift, and above all his unyielding entrepreneurship, he generally acquired both of these in the course of time." It is no wonder that it was the privileged burgher class that opposed the Haugeans, as they were engaging in economical competition. The Danish picture is not as clear; the revivals were mostly a rural phenomenon, with farmers being "the major force" (Ylikangas 2002, 23), and there is a connection between liberating agricultural reforms and the revivals. But tenant farmers and farm laborers also instigated the movement, with peasants joining only later, the majority being young people.

Whether there are substantial differences in the social profiles of the various Finnish revivals has been a debated issue (see Suolinna 1975b, 18–23; Heino 1976, 110; Ylikangas 1979, 308; Suolinna 1982, 98–100; Kinnunen 2004, 166, 356), but the consensus appears to have moved towards seeing the variation as being due more to disparate source materials and base populations than to any inherent differences in the revivalist message. This enables us to consider various explanations for revivalist activity based on the living conditions of various social groups at the time. However, prior to turning to these (in section 7.1) I need to describe the ways in which the revived (who do not appear to have reasoned on the basis of class composition) identified each other and engaged in the creation of new organizatory units in the social environs.

6.2 Self-Organization among the Revived

This subchapter is mostly concerned with demonstrating that individuals who participated in revivals under scrutiny went to considerable lengths in signaling their worries to those around them, and as shown in the previous subchapter, those who responded in kind were likely to have had shared concerns. This led to a convergent process where the revivalists formed cohesive norm groups, and at least in some cases this appears to have been a functional solution to some of the issues that supposedly caused anxiety.

As reviewed above (chapter 5.1), literature on the revivals suggests to me that they are best viewed as a bottom-up, "grassroots" process, where the impetus for action comes not from any single individual or an already established institutional practice but from a multitude of behaviors that need not be homogenous. However, for a recognizable macro-phenomena to take shape, self-organization is required. In this case I refer to the behaviors and practices that resulted in emergence of a community of the revived distinguishable to both insiders and outsiders without any central authority or external element guiding this process. (The role of charismatic leadership is for the most part exluded from the analysis; see chapter 7.2.1 for a discussion of this.)

Drawing on postulates in chapter 1.3, such a self-organizatory process has to involve at least three stages: 1) identifying others with similiar preoccupations; 2) negotiating shared behavioral signals and norms that demarcate the revived from others; and 3) engaging in norm-compliant behavior that preserves the group and is at some level aimed at resolving the original issues.[11] As these proc-

11 Here I reiterate an important qualification: The potential functionalism of the revivals which my account implies is of a crude kind, stemming from a generalized hazard-precaution system, and there is no guarantee that the emergent norm group, however cohesive, can or will grapple with the issues that troubled the participants – especially as it is in outlook concerned with transcendent issues rather than those of this world. A process of selection would be needed to weed out groups that are dysfunctional or whose attempts are misdirected. For an extended

esses are decentralized, they may well overlap, as potentially involved individuals can take different approaches and varying amounts of time in accepting or rejecting what is at stake. They may also fail, especially the third process, in which case the proto-group would dissolve or stagnate.

6.2.1 Behavioral Requirements and Group Demarcation

From the memoirs of Supplicationist leader Matti Paavola (1786–1859) comes a list of tenets for the Jumpers (included in Akiander 1859, 274–276; see also Heino 1976, 25–26). The memoir contains several separatist or sectarian tenets, such as exhortation against non-revived clergy [4, 5, 9], but also instructions to pledge loyalty towards fellow revived [7], visit them often [10], and take godparents only from among the revived [6]. Transgressions against non-revived were not to be recompensed [8]. Lard, hard-to-digest foods, coffee, beer, and spirits were to be avoided [11], and clothing and greetings should make plain that one belongs to the sect [12, 13]. Prostrate prayer was to be always performed when coming to church, to a priest, or to conventicles [14, 15]. In meetings, men and women should perform prostrate prayer lying together in a circle [16], after which there was to be singing and jumping inspired by the Spirit [17]. Akiander (1859, 276) thinks the requirements for prayer, singing, and dancing were the most widely observed, as showing off the gifts of the Spirit was of central importance in this sect.

Ascetism, even self-mortification, was characteristic of the movement (Heino 1976, 129–134). This made the divide between Jumpers and others more pronounced. Active reform was required of those coming to a revival. This took the form of fasting, in addition to the avoidance of the specific foodstuffs mentioned above. Simple and uniform clothing was another form of self-denial (women used white jackets with gathered sleeves and tied a black scarf over their heads, while men grew long beards). In addition to these, prostrate prayer functioned as a sign of being one of the revived. Also some accusations of lewd practices were aired, but these do not stand up to scrutiny (Heino 1976, 143–145). A bishop's visitation on September 23, 1832 (protocol in Akiander 1859, 287–293) confirmed that, "persons, before known for reluctance, sloth and excessive use of strong drinks, after getting into this company, have become sober and hard-working." Heino (1976, 135, 142) thinks the ascetic practices were not held to

discussion of how this kind of functionalism relates to the earlier functionalism of sociological and anthropological tradition, see Wilson (2002, ch. 2).

be an end to themselves or proofs of personal salvation, but to prepare the flesh for attainment of ecstatic experiences, which were taken as a sign of attendance of the spirit and subsequently as a mark of salvation. They, however, did not disappear with the ecstatic practices.

Jakob Wallenberg exhorted his followers to engage in all kinds of norm breaking, both concerning sexual morals and religious conformism (see chapter 5.3.2). But little is known of the doctrine and practices of the movement at Kuortane. It was possibly less radical than at Kauhava, and after court actions an amount of secrecy surrounded the meetings. Rejecting communion, marriage, and the priestly office (Akiander 1860, 35–37) were clearly sectarian practices, but it is untenable to assume these tenets were followed by all sect members consistently throughout the period. As to their other practices, A. O. Törnudds, deacon at Lapua, gave Akiander (1860, 49–50) the following report in 1859, citing "Lisa" as his source:

> In meetings, which were often held at Vasu's house, Isaak first made a speech to elevate the "holy;" thereafter the old man played on his fiddle melodies that were in part sorrowful, in part joyous, or sang Psalms or songs the participants had written themselves; then the whole group went around the hut – "to Heaven." Thereafter all unmarried women took their liquor flasks, which they are all required to have, to the old man, who tasted from each with the effect that he usually became drunk. Those gathered drank the rest. Next was laid out food, which each member was to have brought for "the communal feast of the holy." For this feast every participant must bring his or her part, even if they have to steal it; stealing was permitted, especially for servants. And if someone could not present some delicious morsel, he received reproaches and harsh words; but lard or pork was not allowed, since they were unholy. Liquor was much used, even outside meetings. Everyone should, before he went outdoors in the morning, take a swig of liquor and also constantly use snuff or tobacco, to prevent evil from invading the body.

Other reports corroborate this depiction, also mentioning ritual smoking and ecstatic fainting with visions afterwards; and as with Wallenberg, the dance was intended to stomp on depravity and "the snake" (Akiander 1860, 51, 55; Seppo 1987, 69; Troberg 1995, 49–50). But it is more difficult to assess the truth behind the recurring allegations of promiscuity (Akiander 1860, 42), as the circumstances surely encouraged rumors. Reports claim non-marriage was more common among them, and that they had fewer children than was usual (Kansanen 1915, 152). It is true that their teaching did denigrate earthly marriage, and Isak Vasumäki allegedly stated that illegitimate children were preferable (Akiander 1860, 52). However, the sectarians are not over-represented among those actually accused of promiscuity or giving birth to bastard children, nor do surviving data mark them as anomalous concerning their marital relations (Troberg 1995, 53–56).

As the sectarians already thought themselves to be among the elect on earth, they had cause for continuous elation and praise, as especially evidenced in their cheerful funerary practices (Kansanen 1915, 151; Seppo 1987, 69–70). They also concentrated members' graves in a single corner of the graveyard (Troberg 1995, 50–51), a clear sign of separate group identity.[12] They continued to visit churches and let the priest baptize their children, but they made their reluctance visible, shunning especially the Eucharist (Akiander 1860, 37, 40–41). Despite this they were generally law-abiding and known as helpful people, as long as one did not criticize their ways (Akiander 1860, 38, 56; Troberg 1995, 53).

The Ostrobothnian *Skörts* wore distinctive, plain, and outdated peasant clothing that invited ridicule and also gave them their name (Rosendal 1912, 57–58; Ylikangas 1979, 45, 133, 184, 203). In court, Paavo Ruotsalainen and others explained how they had given up silk and other expensive clothing "inappropriate for the peasant estate" and favored instead something that had been worn by their parents and could be cheaply made at home. While they denied that it was used as a group marker (Akiander 1862, 148–149), the fact remains that it separated them effectively from other people, and with time became a strict requirement among the revived laity. Plain clothing was taken up with such enthusiasm that old, fancier clothing such as silk scarves were donated to fund-raisers for missionary work, as court investigations would reveal (Rosendal 1905, 75; Ylikangas 1979, 44, 220; Hirvonen 2003, 70).

Rosendal (1912, 42, 72–73) states that by the 1840s there was "greatest conformity in doctrine and behavior, from which no deviation was allowed nor attempted," and also "astonishing orderliness" in their life together. In this atmosphere deviations created suspicion and were put down. A memorate describes how *Skört* women in a conventicle, finding blue gloves lying around, cut them to pieces, driving off the lady who had owned them. They are reported to have also given up dancing, gambling, and drinking, and bought and read Bibles and other religious publications more than was customary; devotions and singing were said to be time-consuming, often lasting all night (Rosendal 1905, 383; Ylikangas 1979, 88, 220, 260). During the first court process in Uusikaarlepyy, the prosecutor listed seven illegal conventicles, which revealed that devotional activity for the revived was available weekly, and each Sunday they would travel to those churches where a revived preacher was giving services, even across parish borders and despite attempts by the authorities to stop it (Rosendal 1912, 52, 73; Ylikangas 1979, 55). Many contemporaries testified that the revival had resulted

---

12 One memorate mentions them also separating themselves from outsiders by fanciful dress, but does not specify the period (Akiander 1860, 58).

in a positive change of ways, and even those hostile to them attest to their rejection of amusements and decorations (Ylikangas 1979, 153–155, 260).

As mentioned above (chapter 5.3.3), *"Om sekterismen i trakterna af Kajana"* ("Of sectarianism in parts of Kajana"), a lively though hostile description of the ways of the *Skörts* in Kajaani, was published by the local district physician Elias Lönnrot in *Helsingfors Morgonblad* (49–52, June 29 – July 10, 1835). Lönnrot states, "One recognizes easily, even from the outlook, someone who for some time has been a Pietist; the eyes have a strange glance, deviant from their clear look, the nose as if more pointed, and the cheeks decayed." While this is probably more indicative of his nineteenth century physiognomical expectations, his observation of how one could follow the coattails (*skört*) of still hesitant revived women growing "like grass in springtime" little by little until they were four or six inches long, is intriguing. Barring this, he does not suggest that the changes in their ways were duplicious or false, but rather denies that this kind of self-denial from "dance, music, play, or song" was good or necessary for the soul. He even takes some pains to argue that he doesn't believe those who lost their senses during their conventicles to be faking it, but offers instead an explanation based on scriptural misinterpretation (1 Cor 14) resulting in evocation of psychological needs that spread "like a contagious disease."[13] However, he also openly accused them of sectarianism, as they held for example their dress to be "so important that, according to their faith, scarcely anyone differently dressed would get into the heavenly kingdom."

6.2.2 Emotional Signaling

All three of the revivals under inspection involved ecstatic practices that constitute "emotional signaling." As the name implies, the Jumping Revival was characterized by intense jumping for joy. Akiander (1859, 299–300) includes the following account by a local vicar:

> At the end of divine service songs were sung to a beautiful melody: "Thank you God, Thank you Lord Jesus," and after that "Thank you Jesus for making me one of the elect," which were repeated until all those present began to "jump with spiritual joy" and clap their hands. This jumping continued, with everybody standing in their places until they became bleak, frozen-eyed, and terrible-looking. Exhausted from their exertions, they cast them-

---

13 Lönnrot's attitude appears somewhat typical of the romantically inspired gentry, who saw their ambitions of a national awakening thwarted by the religious revivalists. The latter, for example, gave no value nor aid to efforts in collecting and compiling the partly pre-Christian folklore and poetry of Finnish people (Ylikangas 1979, 222–232).

> selves prostrate on the floor and laid still for a while. Then they stood up and and bade each other friendly good-byes.

Clapping of hands was most common, but conventicles culminated in a bout of prophesying, jumping, dancing, and fainting (Heino 1976, 124–127).

As depicted above (chapters 5.3.2 & 6.2.1), meetings in the Kuortane Revival centered on dancing or marching to music (to "stomp on the snake"). The June 1799 court proceedings enumerate that earlier in the year there had been at least eight separate happenings where people had danced.[14] As the prosecutor demanded fines specifically for those who had danced, it seems that dancing was seen also by him as a group-demarcating feature (Klemetti 1932, 240). Music was specifically written for the purpose, and oft-mentioned is their song *"ompa meilläkin vaunut, joilla taivaaseen mennään"* ("We too have the wagons that will take us to heaven;" Akiander 1860, 55; Kansanen 1915, 151–152).[15] They also supposedly rejoiced constantly at already being among the elect (Seppo 1987, 69–70), and witnesses spoke of convulsions, violent shouting, and visionary experiences at meetings (Klemetti 1932, 236–238; Troberg 1995, 50): "[...] someone among them fainted, at which the others gathered around him, repeatedly blessing. When the unconscious had like this laid for a moment, he stood up and spoke of his visions, and those present received great joy that their brother, who was dead, again came to life" (Akiander 1860, 55).

Rosendal (1905, 370–373) contrasts the "sobriety" and learnedness of Ostrobothnian *Skörts* with the Savonian Revival's emotionalism, and attributes this to a difference in peoples' temperament and the work of the revived preachers. While the former is doubtful, considering the open ecstaticism of various eighteenth century movements in the area (Fräntilä 2001, 184–185), not to mention that of the Kuortane Revival, the latter may well be culpable here. Despite their own emotionalism when preaching, the revived clergymen seem to have frowned upon affective extremities in their flock, so that those engaging in glossolalia during services could be ordered silent or even carried outside (Rosendal 1905, 460, 478–479; Rosendal 1912, 90). In a letter, Malmberg complains of the revived in Gamla Karleby (Kokkola) "sometimes throwing themselves in church, its environs, and in public meetings on their knees and face, sometimes engaging in a fit of solitary dancing, making all kinds of noises, and violently shaking

---

14 Akiander (1860, 39–40) omits the part of these proceedings where dances are discussed, but Klemetti (1932, 236–238) has included a Finnish translation of this part.

15 Here Troberg (1995, 91–92) makes the interesting observation that polska, their dance, came from a wedding tradition, which would mean that participants would associate it also with questions of economy, alliance-formation, and social status.

their bodies, etc. as if every whimsy was given by the heavenly Spirit" (Akiander 1862, 239–240). This is a depiction very similiar to those from the other revivals. It is also significant that in the 1850s, when the revival was in a state of schism after several clergymen had detached themselves from it, ecstaticism surged, possibly because of relaxed control (Rosendal 1915, 78–79). Based on this it seems safe to assume that there was similiar impetus for ecstatic phenomena among the Ostrobothnian *Skörts* as with other revivals, but it was smothered by their ecclesiastic leaders who, despite everything, may have had a deep-rooted appreciation for knowledge of the *word* of God, as evidenced by their enthusiasm in circulating religious literature.[16]

The revived were against what they took to be the depraved lifestyle and nominal Christianity of the gentry, including high-ranking members of the cloth, who had little difficulty in interpreting injunctions against expensive clothing and entertainment as directed against their ways of life (Ylikangas 1979, 155–156; Hirvonen 2003, 38; Peltonen 2006, 231–232). Alfred Kihlman, a *Skört* student, observed city life as consisting of "high words spoken only by virtue of one's office, while engaging in servility, greed, sloth, and lechery," with even members of the clerical estate spending their nights in restaurants and villas (Hirvonen 2003, 26). These "insouciants" (*suruttomat*) appeared to be more interested in gathering earthly possessions than in the fate of their souls.[17] In contrast to this perceived hypocrisy, the *Skörts* emphasized the inability of human beings to achieve salvation through personal deeds and shunned what they held to be self-righteous behavior. The most blatant example of this was the uninhibited drinking practiced by some *Skörts* (Hirvonen 2003, 31–32; but not all, see Rosendal 1905, 388). Paavo Ruotsalainen advocated drinking alcohol on the grounds that it is better to wallow in sin than try in vain to avoid it.[18] Also, their abhorrence of mammon did not prevent the *Skörts* from engaging in lucrative business, and the marriages of their clergy into wealthy families aroused criticism among the nonrevived as being in conflict with what they preached (Hirvonen 2003, 34, 41–42).

---

16 Certainly, such an attitude aptly characterizes Henrik Renqvist, the deacon of Liperi and leader of the contemporary Renqvistian revival, who was a prolific author and translator of devotional literature and strictly rejected ecstatic phenomena (Hartikainen 2005, 268–269).

17 This appears to be a common theme among the revived authors. Even Rosendal (1902, 399) comments quite strongly on the "tiring" habit of Niskanen (1998) to repeatedly admonish against the hypocrisy of non-revived Christians.

18 Rosendal (1912, 70) even claims Paavo Ruotsalainen to have abhorred hypocrisy so much that in some cases his bottle would have contained only water, but that he still pretended public drunkenness – a peculiar act of reversed hypocrisy one need not credit (see also Malmivaara 1914, 5–6, 9).

As already mentioned, the state church of the time was content with parishioners' conformance to the outward forms of Christian practice (such as church attendance and annual communion), with little interest in their spiritual state. That this was enough for the salvation of one's soul was denied by the revived and their clergy, who demanded more from their flock (Rosendal 1905, 80–81; Ylikangas 1979, 120–123, 137–138, 148–152, 162–163). But they also levelled accusations of hypocrisy and improper behavior at their parish ministers, including preferring socializing with other gentry to their religious obligations (Ylikangas 1979, 121, 155–156; Peltonen 2006, 123–126). In their minds, most preachers read rote-learned texts and homilies written by others with little personal commitment (Rosendal 1905, 453–455). These were serious accusations that could have judicial repercussions at the time (Rosendal 1905, 97). In contrast, the preachers from their own ranks showed open emotion with their pastoral duties, and some apparently held their sermons with little preparation, testifying to their personal experience of the Lord's mercy, and threatening parishioners with hellfire (Rosendal 1905, 455–466, 473; 1912, 59):

> It might be expected that I on this first day of the year would wish the congregation happiness, good cheer and fortune, and blessings. But I read people like the signs of workshops in towns. When I come to town, I see several houses with signposts on which are painted boots, scissors, saddles, and whatnot. Each sign tells what trade the craftsman in question practices. Alike are your faces, which bespeak of the prince you all serve. And who is this prince? The Devil himself! And therefore – no blessings, but a curse, a curse over you all! Amen! (Johan Mikael Stenbäck in 1839, quoted in Ylikangas 1979, 136).

This kind of rhetoric and substance appear to have been common with them: accusers claimed that one *Skört* preacher had divided parishioners into a "small group" of believers and a "large group" of nonrevived whom he cursed (Ylikangas 1979, 135–136, 153, 155). One man remembers wondering why, when he was a boy, people always cried in church when the revived preacher was speaking (Rosendal 1905, 464). Ostrobothnian *Skörts* would shun nonrevived members of the clergy and, disregarding injunctions, march on Sundays even across parish borders to hear one of their own preach (Rosendal 1912, 52; also Peltonen 2006, 199). Despite not having priests who fully associated with the revival, the Jumpers practiced similiar discrimination, and contrary to the eighteenth century folk revival in the area (Sulkunen 1999, 22, 71), it was no longer enough for a minister to be diligent in his duties and sympathetic towards the revived. A personal spiritual awakening was also required, and the surest sign of this was open emotionalism (Heino 1976, 158–162).[19]

Did emotionalism become more muted over time? Evidence from the Kuortane Revival is scant, but it seems that ecstatic phenomena were more pronounced in the early stage (Troberg 1995, 58–59). Among Jumpers, the ecstatic displays disappeared after 1830, which coincides with the stagnation of the revival (Heino 1976, 124–127, 165–169). With Ostrobothnian *Skörts* such an appraisal cannot be made, since what emotionalism surfaced was quickly subdued by the clerical leaders. But ecstaticism also grew less in Savonia over time (Ruutu 1967, 99), and in a 1992 newspaper article (Kulpakko [1992] 2007) the vicar of Kittilä describes how the famous Laestadian trances (*liikutukset*) have also been on the wane (see chapter 7.2 on why it has in the Laestadian case taken a considerably longer period of time).

I argued in chapters 2.4, 3.2.3, and 4.5 that on theoretical grounds emotional signaling appears a viable way of ratcheting up trust, especially when other kinds of signals have become too conventionalized to signify anything. Although high levels of emotionalism appear difficult to maintain, new behavioral signals can become established alongside them to mark the cultural borders of a new norm environment that can then be maintained with prosocial punishment (see chapter 2.2.2). However, descriptions of the revivals also tell of other kinds of activities that peaked only in the initial phase and soon waned. From some revivals there are reports of above-the-norm communion activity (Rosendal 1912, 96; Raittila 1976, 61–62, 68–69, 211–214; Kimanen 2008, 35–39, 124–125). That such activity did not become an established institution within the revival movements is understandable; it provoked the ire of not only the clergy but also other parishioners, who would have to pay the costs of increased wine usage (Hartikainen 2005, 234–235; Kimanen 2008, 36). It appears that rather than actively seeking conflict with outsiders, the revived for the most part simply preferred keeping their own company. More difficult to interpret are similiar mentions of revivalism occurring with apparently lessened criminal and litigation proceedings, but this too was short-lived (Rosendal 1905, 83; Raittila 1976, 17, 25; Lohi 1997, 68–69; Kimanen 2008, 81, 151). Within the Laestadian revivals this resulted at least partly from the abstinence strongly advocated by Lars Levi Laestadius and his followers, but the fact that overall levels of litigation returned

19 For descriptions of similiar sentiments among the Laestadians, see Raittila (1976, 206, 235) and Lohi (1997, 658, 665). There is less direct evidence of such emotionalism being a requirement also for laymen, but the Laestadians (with whom the ecstatic moving or *liikutukset* became an institution in itself) had the notion of a "cold jump" (*kylmä hyppy*) for feigned ecstaticism (Raittila 1976, 195–196).

after a few years' dip is more problematic. If this shift was connected to more cohesive relations between (a section of) parishioners, why didn't it last longer? But as these two phenomena appear most characteristic of the Laestadian and Renqvistian revivals, which are not comprehensively analyzed in the present work, answering this question must be left for future research.

6.2.3 Evidence of Norm Observance and Cohesion

In the preceding chapters I have made an elaborate effort towards demonstrating that: 1) cooperation is of utmost importance to individuals, especially in conditions of low environment predictability; 2) on a theoretical level it is well-established that unambigious group-demarcating markers and social norms can uphold a high level of cooperative behavior; 3) in all three revivals under scrutiny, highly visible and costly devices of tracking individual commitment were employed; and 4) revivals found their support disproportionately from social groups where we may expect incentive towards renegotiating ties of mutualism, norm-adherence, and information-exchange. The question remains: is there evidence for the missing link between these considerations, that the revivals were a functional institution that supplied those who joined them with resources they found themselves (however unconsciously) lacking? In the absence of data with which to perform quantitative comparisons, it is hard to establish whether levels of cohesion or norm observance were comparably higher in the revived communities than in the surrounding society. Jumpers' tenets decreed loyalty and relations between coreligionists, to the exclusion of others. The Kuortane sect required unity on threat of punishment and had some practices that indicate norms of sharing and union with their spiritual brethren. The *Skörts* expected hospitality from their fellows and forbade taking interest on loans from them, to mention the most obvious practices, and peer pressure was applied to transgressors. All three revivals include an expectation of indirect reciprocity within the in-group.

As for the evidence, memoirs of the revived (Malmivaara 1914, 72–76) speak of clearing bees and hospitality, but such behaviors were in principle nothing special compared to the general populace. But in some cases the changes in behavior were more striking or qualitative in nature. For example, Kansanen (1915, 153) claims the Kuortane sect to have built "without pay" the house of one coreligionist. The Jumper movement was also unified to the extent that when Luvian Jumpers were severely fined for participation in illegal conventicles, funds to pay these fines were raised collectively, even from neighboring parishes (Rosendal 1902, 240; Heino 1976, 92–93; for a similiar case among the Ostrobothnian *Skörts*, see Rosendal 1912, 56). Some also reported them as being "helpful especially towards their co-religionists" (while also thinking that cheating an unbeliever was no sin), and godparents were chosen from among co-religionists (Heino 1976, 103, 144–145, 161–162). However, instead of listing such diverse examples, my purpose here is to illustrate the forms parochial separatism and everyday reciprocity took in these movements and how they were interconnected.

Some authors have argued that the mere emergence of these sectarian groups meant drastic changes to established social practice (Heikkilä and Seppo 1987, 75–77). Relations might be limited to fellow revived, even to the exclusion of relatives, and a nonrevived farmhand or maid would not find employment in a revived household.[20] A clergyman officiating in a baptism, wedding, or funeral was entitled to a service fee, which could in the case of deacons and chaplains make up a substantial proportion of their earned income (Peltonen 2006, 103). Thus, when the revived favored their "own" ministers (Ylikangas 1979, 19; Heino 1976, 157–160), they were defecting on their parish clergy. In tying salvation to their singular practices they created an exclusive notion of the true church, furthering the divide between them and others (Heino 1976, 162–163). Conversely, the revived clergy offended sensibilities of the gentry by putting their solidarity with the other revived before solidarity with their own estate (Heikkilä and Seppo 1987, 77). Inevitably, such behaviors were to create conflicts, but it appears that merely the large, unified front of the *Skörts* was enough to stop harrassment by less persistent authorities (Rosendal 1912, 52, 54).

Although a norm of group loyalty can manifest as heroic self-sacrifice during open conflict, most often we are likely to observe it as intense but informal networking: small services and sharing strategic information. While such behaviors may not strike us as being very "altruistic" in an everyday sense, their accumulation can be of considerable import, both for the group and the individuals. For example, Troberg (1995, 86–87) considers the social network perspective and notes that already in 1799 the Kuortane Revival was clustered in the heartlands of the parish, and even more so in 1825. Also, comparing the places of birth and places of residence, he finds that there is more dispersion with the former, but the same houses dominate both lists. This is a clear indication that the sectarians sought spouses and employment from within the movement. While this may not sound very surprising, the essence of group solidarity is to direct all kinds of beneficial interactions to those of one's own group. Whether this was the road

---

20 Malmivaara (1914, 41, 84–86) recalls that in his childhood, the division was so strict that children of the revived were not allowed to play with children of the nonrevived. Also according to his memory it was customary among the *Skörts* to take their servants from among their own kind, which he considered important for raising children in their religion. The same went for marriages.

to success is hard to say, but Troberg (1995, 97) states that clearly membership in the movement did not diminish one's standing. He also notes that the collectivity of action within the movement seems to have become even more pronounced with time (Troberg 1995, 60).

The most elaborate evidence that the revivalist activity provided social support can be found in the systematic study of networking among the Ostrobothnian *Skört* clergy, conducted by Hirvonen (2003) through the correspondence of Alfred Kihlman.[21] In his analysis he draws on network theory and the concepts of social, cultural, and symbolic capital as elaborated by Bourdieu (1977, 171–183; 1997) and Coleman (1988) to illustrate the pervasive trust and reciprocity present in their interactions. Bliege Bird and Smith (2005) have argued (with reference to Bourdieu's formulation) that the concept of symbolic capital is actually quite similiar to that of costly signaling.[22] They propose that this connection could even act to unite the "symbolic" and "materialistic" research traditions in the anthropological sciences, but for our purposes it is enough to note that because of this similiarity Hirvonen's method and findings are quite compatible with the aims of the present study.

Hirvonen (2003, 56, 61–64) argues that while networking and exchange of goods, money, and information was characteristic of nineteenth century gentry, the network that formed around the *Skört* clergy was exceptional in its tightness, its high degree of closure,[23] and its constant exchange of services (see also Hirvonen 2000). There were also qualitative differences: instead of amusements (e.g., balls), shared identity was built in conventicles. In correspondence, constant self-mortification and reserving gratification only to spiritual matters worked to uphold a shared system of norms over what was considered proper.

---

21 Kihlman (1825–1904) was one of the Ostrobothnian *Skört* clergy, and later in life a headmaster, a member of parliament, and a business magnate (Hirvonen 2003, 1).

22 Bourdieu (1977, 180, emphasis original) states, "The acquisition of a clientele, even an inherited one, implies considerable *labour* devoted to making and maintaining relations, and also substantial material and symbolic *investments*, in the form of political aid against attack, theft, offence, and insult, or economic aid, which can be very costly, especially in times of scarcity. As well as material wealth, *time* must be invested, for the value of symbolic labour cannot be defined without reference to the time devoted to it, *giving* or *squandering time* being one of the most precious of gifts. It is clear that in such conditions symbolic capital can only be accumulated at the expense of the accumulation of economic capital."

23 "Closure" is here a technical term defined by Coleman (1988, 105–106) as follows: In closed social networks, if *A* knows *B* and *A* knows *C*, *B* also knows *C*. If this holds for a high proportion of triads *A*, *B*, *C* in a network, the network is said to have a high degree of closure. Closure is required for norms to be effective, as in a more open network infractions are more likely to go unpunished due to lack of coordination.

Collective monitoring ensured little deviation from these norms. However, Hirvonen considers the norms to have been for the most part altruistic (prosocial), as they made individuals perform sacrifices for the good of the group, even if this meant demeaning one's position within the estate society or the state church. This interpretation is concordant with the recurring theoretical insight that more effective sanctioning may be one of the key differences between religious and other kinds of institutions in upholding cooperative networks (e.g. Sosis 2005; Chen 2010, 307).

This in-group orientation was also manifest in marriages, organized conventicles, intensive correspondence, frequent visiting even over long distances, and exchanges of services. The *Skört* network allowed the struggling young clergymen to marry above their rank,[24] and deliberate intermarrying created extended families with numerous connecting links. Again, it was not a new thing for the clergy to treat marriages as economic alliances (Karisto, Takala, and Haapola 1997, 153–157), but both Hirvonen (2003, 39–48) and Peltonen (2006, 138, 161) see something exceptional in the *Skört* case. The clergy would also arrange marriages between their parishioners. The common reason offered was that difference in religious opinion would lead to disharmonious couples, but the underlying logic appears to be that the strength of religious emotion and the ultimate concerns it provokes work to create tightly interconnected familial units. When visiting over long distances, *Skörts* could count on receiving lodgings from their brethren, even those previously to them (Hirvonen 2003, 34, 49–54).[25] Although participants in conventicles apparently brought food to be shared (Peltonen 2006, 113), lavish hospitality was often expected, which lead to some near-bankrupties, as poorer *Skörts* also wanted to remain part of the network constructed around extensive visitation. Conventicles, weddings, and funerals could be massive happenings (with hundreds of guests and lasting several days), which offered opportunities for both renewing distant relations and entering into new ones. Consequently, although all feasts included preaching, at one stage the societal functions were seen as usurping the devotional ones, which led to complaints and attempts to "revive" the earlier ideal.

In the absence of a proper banking system, not to mention social security, individuals had to rely on friends and relatives when in need of money. For this the *Skört* network proved invaluable (Hirvonen 2000; 2003, 34, 47–48,

---

24 Their patriarch Malmberg gave Kihlman permission to marry into a wealthy family only on the condition that he would support their poor brethren with his fortune (Hirvonen 2003, 45).
25 Similiarly, a Jehovah's Witness working for the organization told me that even though they receive paltry wages for their work, they can still travel abroad, as in every town there are fellow Witnesses willing to lodge them.

58–61, 64, 74). This function was not limited to the *Skört* gentry, as at the time of his first wife's death, over half of Kihlman's debtors where householders. This required trust, both in potential debtors and those who may have vouched for them, as every new (business) associate meant taking a risk. It is a striking example of in-group prosociality that charging interest for loans was not considered proper between fellow *Skörts* – and equally significant is that when some of the creditors detached themselves from the movement, they started demanding interest also for old loans. Some small loans even went unrecorded, as demonstrations of one's disattachment to worldly things. The letters also contain inquiries of the trustworthiness of third parties and notices on open positions – information of considerable value (cf. Coleman 1988, 104). In an age when the flow of information was slow and unreliable, having friends and relatives in the right place at the right time meant a considerable advantage, for example in applying for better positions.

In the early 1850s, several clergymen detatched themselves from the *Skört* Revival. Ylikangas (1984, chapter 1) has argued this cannot have been due to doctrinal differences (from Beckian influences), but was rather a machinated scheme aimed at securing for themselves a safe return to the state church's fold. Hirvonen (2003, 70–71, 75) thinks the *Skört* network had become too closed and restrictive for the clerical members; its closure (absence of "weak ties," cf. Granovetter 1973) limited their access to outside information. Conincidentally, they also gave up the costly practices, such as the interest-free policy and lavish hospitality. It should be noted that by disengaging as a group (which the machinations Ylikangas attributes to chaplain Carl Gustaf von Essen seem aimed at), they could retain a significant portion of their familial ties and friendships, on which their now more secure position was built, and which apparently lasted a lifetime (Hirvonen 2000).

I devoted several paragraphs to reviewing Hirvonen's (2003) study, since it is hard to see how I could have better demonstrated the ways the revivalists' religious commitments may have opened paths to reinforced mutualism and cohesion. It also points to the importance of religious sanctioning in upholding these norms. Its weight as evidence for my purposes is limited most seriously by the questionable generalization of gentric networks to less literate layers of the revived society. But Rosendal (1912, 45–52, 68–69, 178) also testifies to the existence of similiar practices without clear reference to social class. He states that charging interest for loans was not considered proper between fellow *Skörts*. He describes how *Skörts*' constant traveling to conventicles and to visit each other gave a "peculiar mark" to their heartlands, with masses of people on the move, especially on Sundays, and that when visiting over long distances, they could count on receiving lodgings from their brethren, even those previously unknown to them. He marks how the increase in *Skört* ranks diminished persecution and ridicule and attracted those who were "half-revived" or "unconverted at heart" to their increasingly lavish feasts. Malmivaara (1914, 57–58) also laments how "insouciants" (*suruttomat*) took up wearing the *skört* costume, because they thought it would "improve their credibility in the horse-market." These last two examples serve to illustrate how both the signaling cost of being a sectarian and its information value diminish when the movement grows and becomes more accepted. But taken together, the evidence favors the interpretation that *Skört* practices were successful in creating a tightly interconnected network that exhibited strict adherence to new social norms, both in the costly practices and the heightened levels of cooperative action directed toward fellow members.

7 Discussion and Conclusions

> It should be known that differences of condition among people are the result of the different ways in which they make their living. Social organization enables them to co-operate toward that end and to start with the simple necessities of life, before they get to conveniences and luxuries. (Ibn Khaldûn 1967, 91)

> [...] the purpose of human beings is not only their worldly welfare. This entire world is trifling and futile. It ends in death and annihilation. The purpose of human beings is their religion, which leads them to happiness in the other world. (Ibn Khaldûn 1967, 154)

7.1 Summary: Beyond Sociohistorical Explanations

In the above chapters I have attempted to demonstrate that the human capacity for cooperation requires an explanation that accounts for its extensive scale and historical qualities. General evolutionary mechanisms (kin, reciprocity, or reputation) do not appear to suffice by themselves; instead we must look at how individuals organize themselves into norm-conformant groups, as has been forcefully argued by Henrich (2004). Social norms arise to constrain the variability of individual behavior in tactical situations, and the key analytical concept here was found to be expectability: what is expected of others, what we expect others to expect from us, and how this is influenced by contextual cues.

Following several other researchers, I argued that there is a strong case for a connection between religiosity and prosocial behavior, but in order to observe this connection, we need to be more specific about both terms. "Religiosity" must be operationalized as belief in moralizing, omniscient agents (as argued by Johnson 2005) and demanding ritual practices (as argued by Sosis 2003b) with the resultant prosociality appearing mostly confined to the in-group. But I also suggested that instead of a direct relationship between the two, for which theoretical grounds appear questionable, a more plausible interpretation is that religion empowers norm observance, and whether this leads to prosociality depends on the norms. If the norm system becomes exploitative, emotional expression can serve to validate the authenticity of individual commitment to its transcendent underpinnings.

However, while these considerations lead to a prediction that surges in religious activism should be connected to environmental fluctuation (disappearing resources, cycles in economy, societal tumult), and empirical studies confirm this connection, theoretical explanations for this have been found wanting, not accounting for a functional link. On the psychological level, the recent suggestion by Boyer and Liénard (2006) of a generalized hazard-precaution system that generates ritualized behavior serves to explain how individual perceptions of such threats set the stage for the formation of religious groups. But embedded within the cultural group selection framework, their non-functionalist theory also accounts for adaptive outcomes. On the social level, the classic deprivation theory of Niebuhr (1987) and Glock (1964) was presented as a functional account but could not explain why religious activity is sustained by non-deprived members. The answer, I suggested, lies in the capacity of norms to sustain behavior beyond the original impulses with which they emerged. Taken together, the result is the creation of norm communities that can function as social insurance against prevailing environmental uncertainty.

My case study situated the Christian revivals of eighteenth and nineteenth century Finland in a context of severe societal tumult. Reasons for this tumult included a faster rotation between social classes (resulting from population growth), land reforms that successfully promoted individual enterprise at the expense of traditional communal life, and tensions that arose from the state's attempts to control its subjects, both in the economic and religious spheres, by the mercantilistic policy and a ban on private conventicles. Thus the revivals arose in a context of decreased predictability in everyday social and economic navigation.

The three revivals under scrutiny were found to employ norms that constituted a serious breach with the surrounding society. They engaged in a number of practices that effectively made them stand out from the surrounding populace, creating highly visible communities within which to engage in normative behavior that was more or less prosocial. Their major grievance, the perceived hypocrisy of ordinary Christians – especially that of church clergy – was clearly connected to the ecstatic phenomena that validated the personal spiritual awakening that was required if one was to be considered revived. But it is of specific interest that the revivalists consisted disproportionately of middle-class individuals (namely, householders and craftsmen), who no doubt had shared concerns with regard to their livelihood and (relatively) subjugated position.

It is especially the final point that has also been noted by previous scholars, who have put forth a number of explanations for it. While I suggest that my theoretical framework offers an adequate explanation for this phenomenon, I do not presume to overturn decades of accumulated research. However, it seems to me that while these scholars have been very perceptive as to the causes and symptoms of the revivals, they have lacked a plausible theory which would simultaneously account for all the aspects. Below, I will reiterate my claim: the present framework that functionally combines societal tumult into cohesion-inducing properties of religious revivals via human norm psychology provides cohesion by reconciling most of the apparently conflicting findings and weak points. It also resolves the question of why these individuals engaged in an explicitly religious revival when several of their worries were not (in any direct sense) religious in outlook, and why they were accompanied by such intense ecstaticism (a difficult point for previous accounts). I will also address some specific predictions made in the theoretical chapters, and attempt to clarify the nature of the self-organizatory process in relation to competing theoretical views.

7.1.1 Who Had Cause to Communicate Commitment?

As a natural extension of the signaling hypothesis, Irons (2001, 304–305) predicts that as individuals vary in both their religiosity and need to commit, "*those who have the greatest need to communicate a commitment would be those who are most religious,*" and that as successfully forming committed relations results in individual advantage, "*people tend to use religion to express commitments that are advantageous for themselves*" (emphases his). Those in powerful positions should express religious ideas supporting the status quo, while those in weaker positions are likely to seek change through their religion. Thus, costly signaling should show appropriate variation with gender, class, and other social variables.

Pertaining to gender, results are somewhat mixed. Sources point to women being more drawn to the revivals in their early stages than men; with the Kuortane data this is clear, and also with other Ostrobothnian revivals, but not with the Jumpers. We could interpret the revival there as being at a late stage – from 1831–1832 ecstatic displays were already disappearing – but Ylikangas (1979, 71–72) has arrived at the opposite conclusion (on an unclear basis) to account for the difference in social class. That women should have more cause to seek new commitments than men is not surprising considering their subjugated position in all social groups (Ylikangas 1979, 282–283). But regardless of this, with the passing of time we see equal representation of men and women in every case. It seems that in the end, it was often the whole household that did or did not take up the revival, which is understandable given the disruptive effect a personal revival of one family member could have on familial relations; apparently husbands eventually followed their wives to the conventicles.[1]

---

1 It appears a recurring pattern among Christian movements that women are at the forefront of new, emerging movements, but as the movement stabilizes they are pushed into the background. One hallmark of the eighteenth century folk revival were the female ecstatics and preachers who appear as the initiators or even prophets of the local movements (Sulkunen 1999, 20–22, 66–72, 78–88), as with the later Laestadian Revival, where we also see men subse-

But if the gender imbalance became level over time, social class went the opposite way; over-representation of the householders became stronger the further revivals proceeded. Here one may wonder, shouldn't it always be the poorest and most repressed classes who would be expected to be in need of communicating commitment, instead of the rather secure middle classes? The answer appears to be that when one is trodden too far down, even collectively raising against oppression is out of reach. In his work on country paupers of the period, Haatanen (1968, 2) paints a bleak picture of their potential to better themselves, most aptly captured in an 1846 quotation from J. V. Snellman:

> How can we expect a people to survive, when each and every individual is told: You must sustain yourself, but you may work only on the span of so-and-so many quarter mils, and of the soil on this area you may not seek yield, as it does not belong to you, even though it awaits its tiller. Besides, here is a list of a hundred trades, but you may not choose any one of them. The poor creature who hears this has no property nor education, if one does not count his rags or those rows from the catechism he has learnt to recite without understanding a single one of them. If he is without work or after something better in life, how should he proceed? Society gives him no better answer than this: Become a day laborer for the person who owns the land you are tied to.

The principle of "legal protection" (*laillinen suojelu*) dictated that those without a legal profession had to enlist as lodgers or hirelings into the service of another, or enlist in the military; vagrancy was not tolerated (Haatanen 1968, 28–30, 43, 77). Thus, there was little the lower classes could do to affect their fates – those born in poverty died in poverty.

As has been iterated in chapter 5.2.1, after the enclosure acts the landless had lost their rights to common lands, and many lived on the brink of poverty; nor could they take up crafts, as these were protected (chapter 5.2.2). The situation may not have been as dire with all crofters, and there was regional variation to this (Jutikkala 1958, 371), but these were in turn directly dependent on the householder estate that took over the revivals. Their entry may thus have inhibited the lower estates' enthusiasm towards the movement (Ylikangas 1979, 282; 2002, 17, 26). This finding differs somewhat from (but does not invalidate) the prediction made in chapter four that the observed high socioeconomic status

quently taking control (Lohi 1997, 658). This was also the case with early Christianity (Stark 1997, 99–111), medieval divine and demonic possessions (Caciola 2003, 24), the American Great Awakenings (McLoughlin 1978, 124, 132), and the spread of Pentecostalism in Finland (Helander 1987, 181–195). At present, I cannot offer an adequate explanation for this phenomenon; even though this may partly result from women's subjugated status throughout history, examples from Irons (2001, 304) and Sosis and Ruffle (2003, 714) show substantial context-sensitive variation in the relationship between religiosity and gender within the signaling framework.

of participants in religious movements could result from enhancement due to more intensive cooperation: also the recruitment process may be biased toward those with heightened potential for upward mobility.

Thus we are left with the ecological hypothesis of Ruutu (1967) that it was the land-owning populace who mostly embraced the revivals. This was in turn picked up by Suolinna (1975a, 33–39, 47, 52; 1975c, 67–70), who put forth a comprehensive thesis concerning the revivals. She argues that both the industrial revolution and the "national awakening" of the nineteenth century created conflicts and schism, which she spells out in terms of large entities: towns versus countryside, gentry versus commoners, the propertied classes versus the workers, church versus nation-state, and especially center versus periphery. In support she cites Ruutu's finding and further argues that the revivals were able to create strong bonds between their members by doctrinal emphasis on congregational authority and demarcating markers (dresscode, etc.), while the ecstatic phenomena were necessary to legitimize the claims of the revived in their conflicts with the established clergy (although she also thinks they were "expressions of relief" over banked-up frustration). While I differ with this interpretation mostly on grounds of her methodological approach,[2] there are also empirical problems. Her central argument is that the revival movements were a reaction of the periphery against attempts at control by the center (e.g., their refusal to give up the older translation of the Bible; 1975a, 52–53). But as studies by Ylikangas (2002, 30–35), Raittila (1976, 158–159), and Kinnunen (2004, 166) show, rather than constituting the periphery, the areas where revivals took hold can be more aptly described as "ascendant centers," prosperous and populous villages that were on the provincial level but were less established. Both the tradi-

2 The clearest difference between our approaches is that Suolinna's work can be described as methodologically holistic (as the examples above show), while I draw more on methodological individualism. For example, she speaks of *groups* as having latent and manifest aims, and (despite claims to the contrary in Suolinna 1977, 113) appears overly concerned with the categorization of movements as either social, religious, or political (1975c, 57–58, 63–68; 1975a, 41–46, 53). This is unnecessary if we adopt a more individual-centered view, where a person's motives aren't necessarily demarcated along these lines, and acknowledge that humans can be spurred into action by socioeconomic needs, united into a collective by religious concerns, and as a group enter the political realm. Of course, this polarity of individualism versus collectivism becomes more blurred if we adopt the complex systems view, as these are characterized (among other things) by collectives of agents taking on the appearance of behaving as autonomous agents (see ch. 1.2.1). But imposing such an interpretation on Suolinna's work would be theoretical retrofitting and in any case does not support any inherent division into religious, political, or other entities.

tional centres (such as mother parish villages) and poorer, smaller settlements were cooler towards the revivals.[3]

So far the most empirically convincing account of the ecological hypothesis is that of Ylikangas (1979), who took the insights of Ruutu and Suolinna and combined them with Heino's (1976) innovation of using the conventicle proceedings to chart the social background of the revived. Ylikangas argues that groups strongly associated with the Ostrobothnian *Skört* Revival had discontentment in common (see chapter 5.2). Householders were economically secure but lacking in political rights, as after the liberalizing act of 1789 the transfer of power to St. Petersburg had stagnated all reforms. Craftsmen, though a motley group, had a difficult living due to mercantilistic control and a decline in sales. Of clergymen, it was only the lower clergy, with their insecure future, that felt any attraction towards the revived. Ylikangas strengthens his case by further observing that those householders who joined the revival owned larger than average houses, had more often bought their farms from the crown into a hereditary status, and that locations of manufacturies founded by householders correspond neatly with those of the revival – they were thus the more enterpreneuring individuals among their estate. Of the clergy, almost all were themselves sons of clergymen, some in a long line. They had failed to fulfil both their own and their families' expectations, no doubt creating frustrations. Becoming a minister for the revived would have given these men a feeling of being needed, and even of prestige, and occasioned condemning the gentric lifestyle they could not achieve in any case.[4]

---

3 Although they exceed both the temporal and cultural bounds of the present study, Suolinna's interpretations of the functions of twentieth century Laestadianism in northern village communities are intriguing and fit with the signaling hypothesis: "In a village community mutual dependence is of great importance, so in practice it is the only option [to accept your neighbor as he is]. In Laestadianism, the acceptance of your fellow man, or neighbor, is religiously sanctioned. When sins are forgiven within the Laestadian congregation, this confirms also that men, or neighbors, accept each other. The conventicle institution can even be used to repair dysfunctional relationships between neighbors" (Suolinna 1977, 114, translation mine). This unity could then be employed to defend against the "hostile" outside world and to attend to the interests of its members (who are still mainly farmers and business operators).

4 This is in striking parallel to the above-mentioned observation that it was in the ascendant population centers that the revival took hold – (perceived) potential for upward mobility seems to have been a precondition for a general acceptance of the revival (Ylikangas 2002, 30–35). More problematically, Ylikangas (2002, 20) attempts to explain the social differences between the Jumping Revival and the Ostrobothnian case by referring to the largely non-charismatic nature of the latter, and arguing that it was open emotionalism that made the clergy, and subsequently the middle classes, shun the former. While plausible, existence of such a dynamic is not supported by the data from Kuortane.

The explanation offered by Troberg (1995, 22–33, 78–81, 87–89, 95–96) for the Kuortane movement is quite similiar to this (even though he appears unaware of Ylikangas' work); Kuortane had in the latter half of the eighteenth century experienced an economic boom from tar business and new settlements, doubling the number of houses from 1723 to 1808. But by the turn of the century no more clearable land was available, and a traveller's report tells of clearly over-exploited tar forests. The populous generations from 1770 to 1800 could not go on living like their parents had, but alternative subsistence strategies were hard to come by. This resulted in farms being split, sometimes until they were unviable, so children of householders also ended up as crofters and hirelings. Between 1805 and 1825, local lawsuits were mostly about these splits, inheritances (that rarely went uncontested), debts (even small), and land. Economic success resulting from individual enterprise had become the primary measure for a man, but this went badly with the fact that there were limited opportunities available even for the enterpreneurial. This dilemma is likely to have especially vexed children of the oldest estates, those at the forefront of the boom and the new ethos, and now the first to be split among siblings. It is exactly what Troberg found: the revival focused around households that originated from the older echelons of original estates that had been split at least once. Individuals with this background may have experienced deprivation, if the possibilities open to them did not match their expectations or the spirit of success and survival they had internalized at home. Further corroboration is offered by the fact that over the observed time period the movement failed to attract new members except among the same generation most of the original members had come from.

But here the ecological hypothesis hits a dead end: why did the socioeconomical frustrations and deprivations of these groups erupt as a *religious* revival? Ylikangas (1979, 282–285) attempts to tackle the issue by suggesting that it was because the political aims of these people were not yet fully consciously realized, and for this reason their protest against the state bureacracy and complacent higher estates took the form of politically unfocused religious separatism. This conclusion echoes the deprivation theory of Glock (1964, 151) that a shared feeling of deprivation results in the formation of religious groups mainly when "the nature of the deprivation is inaccurately perceived or where those experiencing the deprivation are not in a position to work directly at eliminating the causes." However, as seen above (chapter 4.1), this theory is by itself insufficient. Also, it attempts to explain away the religious dimension in order to make sociological theories and concepts more applicable, which disregards the fact that the revivals were explicitly religious movements, that the revived themselves put their experiences and motivations in religious terms, and that *vis-à-vis* their supposed sociopolitical and economical motives their restrictive and ruinous religious behaviors appear misguided and counterproductive (cf. Heino 1977, 35–36).

If we reject the purely sociological explanations, how are we to account for the empirical findings these authors have employed in their support? My argument is that the theory of ritual signaling and group cohesion works to unify not only these but also other scattered explanations offered to account for phenomena surrounding the revivals. First, within it, they can be intepreted as a class-determined protest and a statement of unity against out-group members perceived as hostile or parasitic (cf. Ylikangas 1979), while enabling and tightening networked cooperation among the in-group (cf. Hirvonen 2003). Their occurrence at times of relative deprivation (cf. Troberg 1995; Sulkunen 1999) becomes understandable when we synthesize theoretical and empirical evidence to suggest that such experience works to increase the urgency of religious expression, which in turn functions to attract other individuals with similiar preoccupations. The idea that religious motivations can succeed in upholding unity where others fail is supported by empirical work (Fishman and Goldschmidt 1990; Sosis and Bressler 2003; Sosis and Ruffle 2003; Chen 2010) and is also theoretically plausible (see chapter 3.1.4).

Second, a theme emphasized by older works in church history and by the revived themselves is the sincerity of revived religious emotion, the burning concern over their spiritual state, and how these arose in the context of the perceived hypocrisy of nonrevived Christians and the worldliness of the established church (cf. Niskanen 1998; Rosendal 1902, 1905, 1912; Heino 1976).[5] The sociological theorists' arguments of latent self-interest guiding their behavior have met with criticism (Heino 1977; Raittila 1985a; Ylikangas 1985; Raittila 1985b). Here the insight of Schloss (2008) that emotional signaling can solve the problem of "Machiavellian signalers" who only go through the motions in order to receive the benefits (see chapter 6.2.2) makes it possible to unite the two conflicting interpretations. In an environment where professing the outward forms of Christianity gave little indication of the norms an individual was committed to, it was precisely the deep emotional sincerity of the revived that made it possible for them to band together. This in turn provided a solution to their other, more material concerns. I find it important that the present hypothesis is the only theoretical construct that actually *predicts* the appearance of ecstatic phenomena under these circumstances. Previous explanatory approaches have seen them

5 Kimanen (2008, 176) writes (my translation): "If the revived were allowed to explain their revival themselves, as they in some cases did, to the authorities, they would consistently state concern for the salvation of their soul and that of the others as the reason."

variously as a form of lay-legitimization (Suolinna 1975c; Ylikangas 1979), a way of releasing pent-up frustrations (Suolinna 1975a, 47; Siltala 1992, 42–49), or even a shamanistic remnant (Sulkunen 1999).[6]

And finally, if we interpret the signaling hypothesis within an adaptive systems framework there is little need to create dichotomies between the "original," religious function of a social network and the socioeconomical character it inevitably seems to take on (cf. Hirvonen 2003, 62). Dichotomies between political and "pre-political" movements are also unneccesary (Suolinna 1975a, 41–46, 53; Ylikangas 2002, 29), as these questions become problematic only if we postulate a division between political, religious, and some other kinds of movements and attempt to reduce one kind to another. In view of the present argument, that religion functions to create cohesive social bodies to address whatever hazards individuals perceive, it is only natural that these unions (if they succeed) subsequently apply themselves in enhancing the position of the in-group in all relevant arenas. It is thus no wonder that revivalists also came to exert openly political influence (Siltala 1992, 417; Huhta 2007, 224–227).

The present view may also shed some light on why the earlier eighteenth century revivals appear to us as phenomenona somewhat distinct from those that occurred around the turn of the century. These included the mystic and separatistic revivals of Ostrobothnia – the most notable being that of the Eriksson brothers and those in southwestern Finland best known for the preaching of Anna Lagerblad, Juliana Söderborg, and Anna Rogel, among others (see chapter 5.1). But they all appear to have quietly died down. It wasn't until the nineteenth century that the revivals took hold and consolidated as wide-ranging movements. The four contemporary revival movements – Supplicationism,[7] Laestadianism (in its various forms), *Skörts*, and the Evangelical movement – originate from this period. So why didn't the earlier folk revivals last? For one, the authorities appear to have been more successful in taking matters into their hands: The

---

6 Siltala (1992, 81–118) argues the revived leaders had also suffered from anxiety, distress, and even obsessive thoughts before they found release in religious pursuits. While his description evokes that of Boyer and Liénard (2006, 601) regarding "an arousal state in which non-action is intuitively considered dangerous" (see ch. 4.2), his psychoanalytic approach may not strictly cohere with the cognitive-neuroscientific framework of the latter. However, Toivanen (1997, 51) also considers the distress of Merikarvia householders over enclosure processes to have been "real."

7 It is customary to date the origin of Supplicationism from the revival begun by Liisa Eerikintytär in 1756 (Haavio 1965, 29; Salomäki 2010, 37), but even if some continuity of tradition is acknowledged, the Jumping revival of 1817–1836 appears as the stage during which a geographically and socially united Supplicationist movement with distinct doctrinal views was created (Heino 1976, 228). See ch. 5, footnote 17.

Eriksson brothers were sentenced to exile, Söderborg was admonished by a court, and the Merikarvia Revival that sprung up around Rogel received royal attention for fear of it spreading (Ylikangas 1979, 19; Sulkunen 1999, 69–77).[8] But what I suggest separates the two periods of revivals from each other is the 1789 act that granted householders full rights to their lands. As has been elaborated in chapter 5.2.1, it was (together with the slow-moving enclosures) a drastic change to living conditions for these people who were now considerably more secure from the pressure and machinations of the higher estates, although still politically subjugated. County constables and churchmen who later tried to prevent the revived from having their conventicles ran into an assertive front of householders who knew their worth. The situation would have been different prior to the act, as obstinate parishioners could be bullied back into line by an alliance of civil servants who were entitled to their services, the gentry who could oust them of their lands, the burghers to whom they were forced to sell their produce, or the clergy who monitored their lives. This would also account for the lower classes' diminishing participation in the later revivals, as (due to developments outlined in chapter 5.2) they had by this time even less security, and their interests would have conflicted with the now-powerful householders.

7.1.2 Was Joining a Revival Costly?

Several of the practices of the revived (see chapter 6.2.1) correspond to what Sosis and Bressler (2003, 219–220) operationalized as "costly" behaviors (such as requirements on clothing, sexual relations, and constraints on gambling or foodstuffs) that entail inefficient accomplishment of an individual's somatic or reproductive goals either directly or by limiting an individual's possibilities of socializing with outsiders. But some might question whether a ban on smoking, gambling, or alcohol is actually costly in terms of biological fitness, since these practices would seem to be mostly detrimental to our somatic wellbeing. They might also question whether the alleged promiscuity of the revived at Kuortane involves potentially siring and bearing offspring and is thus an evo-

8 The Rogel revival is also noteworthy for how it demonstrates some of the hallmarks of the later revivals, as seen in the work of Toivanen (1997, 44–50, 62, 65, 72, 83–85). Parishioners pitted themselves against the vicar, whose personal piety was highly questionable; the context of accelerating use of forest resources for personal gain was an issue; and thehouseholders attempted to present a unified front to prevent the landless from "stealing" the common lands from them by settlement.

lutionary benefit, not a cost.[9] But there is little need to go into these discussions here, as several aspects of the revivals make plain that the partakers were risking substantial financial, physiological, and other kinds of costs.

Most immediate of these were the social stigma (especially in the early stage) and the very real punishments that joining a separatist congregation invited. Some contemporaries thought the *Skörts* purposefully invited persecution with their odd garments in order to achieve a degree of martyrdom (Hirvonen 2003, 22). While when pressed Paavo Ruotsalainen and others gave some economic justifications for their clothing (Akiander 1862, 148–149), it is plain from the outsiders' frequent diatribes that these considerations hardly diminished the overall costliness of this practice. In contemporary Norrland, luxury goods such as silk were in high supply, so that householders could dress themselves and their children as finely as the local state officials did (Ericsson and Harnesk 1994, 46); the high amount of silk clothing that turned up in the *Skört* fundraisers (Ylikangas 1979, 44) testifies to the similiarity of the situation across the Bothnian Bay. However, the authorities appeared concerned with poorer people donating their most valuable possessions to these fundraisers (Rosendal 1905, 75–76).

There are other examples that speak of the personal sacrifices involved in joining a revival. Among the *Skört* clergy, worldly careerism was frowned upon, and they customarily forewent applying for higher posts or degrees. They also submitted to harsh public criticism from their elders and dictates concerning their marriages (Hirvonen 2003, 36–38, 44). But most significantly, as has been rehearsed several times, the revived were summoned to court to answer for their religious practices. Even though the heavy punishments meted out by lower courts were in many cases overturned by higher appeals (Ylikangas 1979, 193–194), there is no way the revived could have foreseen this. The progressively increasing fines enacted by the conventicle placard spelled financial ruin, and a prison term could also mean ruination of one's health.[10]

---

9 Here I could argue that in *ceteris paribus* there is no fitness benefit in having illegitimate over legitimate children, while the social stigma and punishments diminish the benefits of the former. For a similiar and more illustrative argument see Kitcher and Vickers (2003, 151–152).

10 The placard threatened conventicle organizers with a fine of 200 silver thalers or 14 prison days for a first offence, 400 thalers or 21 days for a second, and two years' exile for a third. For mere participation, the fines were 40, 80, and 120 thalers progressively. For comparison, 28 days in prison was considered the maximum punishment, since a longer period of water, bread, and confinement would in practice mean a death penalty (Ylikangas 1979, 15–16, 95). A monetary index is more difficult to provide, but in 1800 the average total wealth of a town populace ranged (depending on the town) from 23 to 202 *riksdaler banco* (in value, roughly equal to silver thalers; see Hirvelä 2009), although the Ostrobothnian towns were noticeably richer than others, the same being true for local peasants (Jutikkala 1949, 185, 195). However, in practice the fines

There is little doubt that partaking in a revival was in itself a costly act, and that those who did so were risking their social status, their fortune, and their health. Such behavior is understandable only if they sincerely believed that the fate of their immortal souls was at stake and valued this above all other earthly prospects.

7.1.3 Were Conventicles Rituals?

Theoretical development of the ritual signaling theory has not been limited to ritual *per se* but has also discussed beliefs and group markers (Sosis 2006). But I find some tension in how "ritual" is treated as a theoretical construct and how it has been operationalized in empirical studies. In their commune study, Sosis and Bressler (2003, 219–221) simply counted all kinds of demanding practices regardless of whether they were ritualized or not. And in some later studies (e.g., Bulbulia and Mahoney 2008) any kind of costly act performed in the name of the group has appeared to entice ingroup solidarity. Theoretically Sosis (2003b; Sosis and Bressler 2003, 227–230) has followed Rappaport (1979a, 1999) in arguing that it is the features of rituals which repel ambiguity (such as their multimodality; cf. Rowe 1999) that communicate commitment, and my discussion in chapter 3.2.2 came to a similiar conclusion. Of the tests on signaling theory, the *kibbutzim* studies (Sosis and Ruffle 2003, 2004) hit closer to this mark: men who visited the synagogue daily were most trusting of each other. Also, studies by Soler (2008) as well as Ginges, Hansen, and Norenzayan (2009) concluded that it is specifically participation in public encounters with religion that fosters cohesion. For this reason, it is important to see whether the Finnish revivals included behaviors that can be characterized as public rituals in this sense.

Within the Finnish revivals the clearest example of rituals were the conventicles, which were the hallmark of revivalist activity. At the same time, their illegality made mere participation a costly act. In terms of Rappaport's (1979a, 175–179) "obvious aspects," they were clearly *performances* in which members of the congregation participated in earnest. But their *formality,* communicativity, or instrumentality is not as apparent. Descriptions of the Kuortane conventicles have an air of raucousness with little structure in them. Still, "Lisa's" description (Akiander 1860, 49–50) indicates the recurrence of a pattern over multiple meet-

meted out were sometimes considerably in excess of those imposed by the placard (Ylikangas 1979, 47–48, 95–96, 145–146, 185).

ings; the liquor sharing and "the feast of the holy" have an especially formalistic bent to them. Their dances would have been synchronous activities in the sense meant by Wiltermuth and Heath (2009), and the same applies to the Jumpers; the text in *Uusi Suometar* #235 speaks of dancing and hollering until rapture and exhaustion overtook them, but Paavola's memoirs (Akiander 1859, 274–276) also mention prostrate prayer as a requirement for entering the conventicle, with men and women lying together in a circle during the more inspired outings. The *Skört* conventicles appear to have been more sober and akin to church services, no doubt due to the fact that they were often led by clergymen who discouraged ecstaticism (see chapter 6.2.2). One witness mentions that they were ardent singers, so that when one hymn was over a new one was begun immediately and spontaneously, without announcement (Rosendal 1905, 96; we can take his word for this, as such behavior was distinctive of their meetings a hundred years later; Haavio 1965, 41–42). The county constable's letter described above (chapter 5.3.3; Rosendal 1905, 90) stated that they held their devotions to be so important that they did not allow them to be interrupted, "even if the roof was on fire," which can be taken to indicate that the assembly held a sacred, or at least inviolable, quality.

While on a continuum of formality these events fall short of being "liturgical orders" with invariant sequences of stylized utterances, that is perhaps to be expected with newly emergent ritualistic phenomena; the tendency of rituals to become more and more fixed over time has not yet affected these. Rappaport's espousal seems to take the existence of rituals as a given and does not cover such diachronic processes. As to their communicativity or instrumentality, this is clearest among the Jumpers; they sought communications from the Spirit in order to demarcate between those who are among the elect (Heino 1976, 119–120), which in effect "serves to express the individual's status in the structural system in which he finds himself for the time being" (Rappaport 1979a, 179, quoting Leach 1954, 10–11). What is known of the beliefs of the Kuortane sect (Akiander 1860, 35–37) points to a similiar interpretation, with their emphasis on "the elect" (who were supposed to rejoice even in funerals of their fellow revived) and channeling of the Spirit to distinguish between the godly and ungodly. With the *Skörts* such interpretations are not as easy to make. They were accused of considering themselves "enlightened" or "newborn" after their conventicles, but in their apologias they are careful to rebut the accusations of factionalism, and instead point out how these devotions present a more Godly alternative to their former (and others' current) ways of drinking, dancing, gambling, and brawling (e. g., the defense of Paavo Ruotsalainen at Kalajoki, included in Krook and Salomies 1937, 54–56). Communicating the word of God so that hearers would embrace God's mercy and righteousness instead of worldly good and awaken to experience in their hearts and consciences the sinfulness of their past ways appears as the explicit *raison d'être* for having these services (Rosendal 1905, 111).

It is clear that the revivals' self-organization was dependent also on other behavioral requirements besides conventicle attendance; both *Skörts* and Jumpers had a dress code and restrictions on food or activities. From Kuortane we do not know of such requirements being widely observed, but there is one crucial difference between it and the two others: the doctrine of Wallenberg and Vasu never spread from its points of origin, while the *Skört* message spanned huge tracts of land, and Jumpers were apparently well connected with their co-religionists from other parishes fifty or even hundred kilometers away (Heino 1976, 78, 83, 92–93). It is conceivable that at Kuortane, where everybody could recognize each other, the frequent conventicles could suffuse the congregation with a sense of community even without constantly observable markers. But the situation would be different with strangers claiming to be fellow believers who came from other, even culturally different, parts of the country. With them, constantly observable badges made allegiances clear to both insiders and outsiders; one could neither walk into a conventicle with silk gloves without attracting unwanted attention, nor go around in an outmoded dress without cries of derision from bands of boys (cf. Rosendal 1912, 72–73). Strict chronological development of these practices cannot be established, but the spread of the *Skört* dress from Savonia appears to coincide roughly with the revival gaining momentum as a unified movement (Ylikangas 1979, 201–204).

In fact, this empirical finding follows closely the analytical argument put forth by Sosis (2005) that in closed religious communities there is little need for trust,[11] because institutional mechanisms of punishment and reputation suffice to enhance cooperation. In Kuortane, according to A. O. Törnudd, "none of the forbidden deeds in God's law were held really unclean or punishable, but moreso transgressions against the sect's injunctions" (Akiander 1860, 50). Not so in more open communities, as multiple identities and intergroup mobility raise monitoring costs and make sanctioning less effective. Here badges and symbolically marked practices are needed, preferably ones that are hard to fake. Neither is emotional identification with every member possible, calling for a rise in the frequency of ritualized displays (Pyysiäinen 2009, 38–39; cf.

---

11 Trust is here defined, following Bacharach and Gambetta (2001), as a situation where A expects (trusts) B to do X even when B's not doing X is damaging to A and strictly beneficient to B. For example, a tourist trusts a stranger to carry his bags to the hotel for a paltry sum instead of running away with them. This definition excludes situations were threat of sanctions makes cooperative behavior preferrable to both parties.

Whitehouse 2000). In fact, a simulation experiment by Macy and Skvoretz (1998) demonstrates that after stable cooperation has emerged in a small community, individuals are also willing to trust strangers who carry similiar cultural markers. To me, this appears to be a remarkable convergence of adaptive complexity between empirical and simulated systems.

7.1.4 Was the God of the Revived a Moralist?

In the theoretical chapters above, a central prediction was that the cohesiveness of religious movements results not only from costly signals of commitment but also from the belief of being constantly under the surveillance of a moralizing deity. Of the three revivals under scrutiny, both *Skörts* and the Jumpers believed in a God that was concerned with the human condition, although one could not be certain that even proper behavior brought salvation. The *Skörts* appear to have put more emphasis on the sinfulness of immoral behavior, while Jumpers tried to prepare themselves for an ecstatic communion with God. But compared to ordinary Lutherans who were assured of being good enough Christians if they participated in set rituals and refrained from criminal acts, the revived lived under constant doubt as to the fate of their souls. The situation appears different, again, with Kuortane Revival, whose doctrine was more vague regarding what was actually forbidden, and who in practice seem also to have had more licence.

It may be a coincidence that of the three, only the Kuortane Revival was short-lived, with the group dismantling after a generation, but it is in keeping with the prediction that norms, especially in large, diffuse groups, are better observed when enforced by beliefs in moralizing deities. And, as has been mentioned, the *Skört* and Jumping revivals spread into several parishes, whereas the Kuortane community remained local.

7.1.5 Finnish Revivals as Cultural Group Selection

Above I attempted to demonstrate how several features of the Finnish revivals correspond to what have been operationalized as the cohesion-inducing properties of religious systems, and that some of the predictions made by the signaling and monitoring theorists are also borne out: individuals who partook in revivals came mainly from social groups that experienced relative deprivation and engaged in costly ritual performances under the eye of a watchful God. But even the three cases analyzed do not appear equal. While arguably costly, the Kuortane revivalists' notion of God was not strictly moralist, and their group markers were less visible, so they lacked features that are in theory associated with cohesive, clearly demarcated norm groups. In terms of longevity and diffusion, it was also the least successful of the three. However, the strength of this argument is weak, as we are dealing with non-deterministic phenomena, and the small sample size does not allow for probabilistic arguments. At best, it does not conflict with the theory.

However, there are two other aspects of the revivals that are of special interest to the cultural group selection hypothesis. First, as has been iterated in chapter 2.2.2, selection in cultural groups can withstand the effects of between-group migration if a group's norms are enforced on those who join it. Consequently, the group identity and its cultural makeup can take on a life of their own regardless of the human constituents (as long as they are capable cultural learners). This is evidenced to a surprising degree by the Kuortane Revival, whose membership changed almost completely in 25 years, while retaining its original pull to those of specific generational cohort. This finding can also be analyzed in terms of the individual signaling hypothesis, that the revivalists signalled their group commitment through norm observance, and those whose commitment dissipated ceased to be counted among their number, since they no longer partook in the normative ritual activity. Within this model it is less clear how the norm system in itself stabilized to provide such a framework for commitment. Also Sosis and Bulbulia (2011, 347) find the behavioral ecology model lacking in this respect and predict convergence with theories of cultural evolution that will remedy the issue.

Second, of the possible forms of between-group competition, Henrich (2004, 28–29) predicts that prestige-biased selection, where a successful group's norms are observed and copied by outsiders, is the fastest form of selection, working on a time scale of decades. This is what happened with the revivalist groups, as by the late nineteenth century the diminishing conflict between them and the authorities had resulted in the revivals overwhelming the church from the inside (Juva 1962, 191–194). Most prominent in this are the *Skörts* (Niemelä and Salomäki 2006, 363–366). While 16 percent of contemporary state church employees say they belong in the movement, another 27 percent describe their thinking to have been influenced by it. Additionally, 81 percent of vicars belong to or have background in some revival movement[12] – a stunning number compared to the 11 percent claimed by the general populace. Within the church, the beliefs and practi-

12 It must be noted that in these figures is included also a smaller number of adherents to movements that do not originate from the nineteenth century revivalism, for example twentieth century charismatic movements.

ces of the revived have proven more successful than those promoted by mainstream Christianity.

In conclusion, turn-of-the-nineteenth century Finnish revivalism arose from the legal and economic changes that put into motion a process of upheaval in contemporary society, creating shared experiences of deprivation and uncertainty across those social groups that were most affected by it. This anxiety motivated ritualized behavior that, being widespread, aggregated into ritual groups, giving birth to conventicles and dress codes. As the church establishment appeared to be a hypocritical defender of the old order, an element of ecstaticism was necessary to differentiate between those who professed belief in order to profit and those whose commitments were motivated by a more sincere experience of the divine. This turned them into separatistic congregations, with the overriding purpose of saving one's soul. But once established, these groups were also in a position to work on the socioeconomic deprivations of their members, with social norms that directed their prosocial behavior towards the in-group and encouraged a united stance against the out-group. Both the interpretation of the evidence and theoretical analysis moderately favor the cultural group selection view of the evolution of religion, with rituals and deific beliefs acting to enforce group boundaries and internal uniformity. Both also emphasize the importance of considering both the socioeconomic environment and norm systems when trying to account for dynamics of religious groups.

7.2 Future Research

The present study has been exploratory in nature, its central aims being to demonstrate the viability of signaling and complex systems approaches in accounting for revivalist phenomena. As such, it opens up several new avenues in which research is needed to more decisively show the validity of the present conclusions.

7.2.1 Reinvestigating the Role of Religious Leadership

The perspectives considered in the present work have largely bypassed questions of leadership that have traditionally been important to many other approaches (including antecedent studies on Finnish revivals, e.g., Huhta 2007, 27; Siltala 1992; Remes 1995). This exclusion is perhaps best viewed as arising from the epistemological strategy favored (what phenomena can be shown to emerge without resorting to design by an individual genius) rather than from inherent problems in reconciling these views. For example, a simulation study on emergence of complex polities by Gavrilets, Anderson, and Turchin (2010, 70) testifies to the importance of leadership in holding together a society.[13] In game theory, a class of problems known as coordination problems equates leadership with a willingness to act first and decide the specifics of coordination, which may also involve competition for this position given conflicting interests (King, Johnson, and van Vugt 2009).

Regarding religious leaders, in an early version of the signaling hypothesis, Cronk (1994, 92) put emphasis on the manipulative, asymmetric use of signals: those who make the loudest proclamations of need for ideological commitment are trying to make *others* commit to these ideals so they can benefit. But mutualism and exploitation can live side by side in a single tradition, and in the above chapters I have argued that ecstatic revivals in part result from human sensitivity to hypocritical professions of belief without accompanying deeds. This is perhaps especially true on the part of (religious) leaders, as people pay more attention to what prestigious individuals are saying and doing, as predicted by Henrich (2009, 257). In an interesting extension of his model, Wildman and Sosis (2011) found in an agent-based simulation experiment that in addition to costliness, group survival was linked to consistency between a leader's beliefs and behavior. Charisma, operationalized as an amplifier of leader effects, had only a minor role to play (but here there is certainly room for other operationalizations also).

However, as leadership is avowedly important for groups, its effects are surely not limited to those stemming from the need to control leader excesses. Hooper, Kaplan, and Boone (2010, 634) predict that transition to cooperation under leadership will occur "when the benefits from being able to successfully cooperate at larger scales exceed the costs of having a leader" and construct a model that shows this is viable on a larger parameter region than mere mutual monitoring (although they can plausibly interact to form complex societies). Within the religious sphere, accounts on the early histories of Christianity and Islam point to the role of leaders in creating punctuation to the otherwise amorphous process of tradition formation, by pointing out to their followers where meaningful distinctions (should) lie. Early church fathers agonized over believers in Christ practicing their faith in synagogues and following Jewish ways (Boyarin

---

13 "The chief's expected time in power τ is one of the two most important parameters. This finding strongly supports arguments on the crucial importance of having well-defined and legitimate mechanisms of succession for the stability of polities" (Gavrilets, Anderson, and Turchin 2010, 70; the other important parameter controlled the amount of randomness in determining how likely the stronger polity is to overcome in a conflict).

1999, 11, 17) or getting mixed up with Valentinian "gnostics," who to lay people appeared in speech and ritual not unlike other Christians (Dunderberg 2008, 2–3). According to Donner (2010, 203, 217, 221–223), early Islam separated itself from surrounding monotheisms only by the time of the fifth *umayyad* caliph 'Abd al-Malik (late seventh century), who introduced a polemic against the idea of the Trinity and reasserted the singular prophethood of Muhammad. My suggestion is that in order to create cohesive groups there is a need for leaders' drawing such salient demarcations, so those who wish to be their adherents can unambigiously differentiate "us" from "them" (one system from another). Especially in the creation of a new religion (see chapter 4), boundaries between identities are often fluid and can in localized interactions be negotiated in several ways, especially on the group periphery (e. g., is his one true God who decrees abstinence the same as my God who requires monogamy?). In a minimal sense, in decreeing proper ritual forms, the leader acts first and decides the specifics of coordination for others to follow.

I am not saying that group organization is dependent on a central authority figure, but rather that their function appears to be that of further encapsulating the proto-group *vis-à-vis* neighboring groups. This certainly applies to many of the leading figures in the revivals under scrutiny. Paavo Ruotsalainen clashed powerfully with the clerical leader of a neighboring Karelian movement, Henrik Renqvist, over the questions of prostrate prayer, temperance, and ecstatic displays, with Renqvist rejecting ecstatic displays but defending the other two; both forbidding their followers from having anything to do with the other party, despite some close affinities and few theological differences (Hartikainen 2005, 269–273). Nor were relations good between Ruotsalainen and the Southern Savonian "Friends Movement" leader Margareta Högman (Hökkä 2006, 53–64). But in Ostrobothnia clerical leaders chose to bow to Ruotsalainen, precluding inter-group hostility and creating a vast, unified *Skört* movement (Rosendal 1902, 290–299; Ylikangas 1979, 234–239).

A more complete theory of religious groups and revivals requires convergence between theories that concentrate on the populational level and leader-oriented histories. Challenges to this include the intermingled factors driving bottom-up demands for leadership and exploitative social dominance (e. g., management of diffuse resources; Hooper, Kaplan, and Boone 2010, 644–645). The hypothesized leader-effect on the centrifugal tendencies of group periphery could be tested with simulation experiments. Another possible avenue is exploring the relationship between leader personalities (often recorded in historical sources) and the context in which they arose. Recent social dilemma experiments by Halevy et al. (2011) show dominance and prestige (two components of status) to be determined by available courses of action and choices made, which should differentially influence leader election so that in times of intergroup peace those unnecessarily harmful to outsiders (dominant individuals) are not favored.

7.2.2 Simulation Experiments and Religions as Complex Adaptive Systems

Whereas the study includes only one simulation model, more could be crafted to test quasi-experimentally some of the more adventurous hypotheses for which purely empirical validation is hard to perform. One is the theoretical suggestion that emotional signaling can work to stabilize norm observance based on costly signals. This could be done via genetic algorithms (as in Macy and Skvoretz 1998), in which emotions are operationalized as a strictly reliable signal of norm observance that are however less stable (i. e., agents are more likely to mutate into a non-emotional state).

More complex is the observation, made on the empirical cases, that the social composition of the revived developed towards dominance by middle-class groups. One individual-level mechanism that might create such a dynamic is that an incentive to signal adherence to the revival is related inversely to one's standing, but individuals choose with whom to interact based on both the other's signal strength and standing. The most desirable partners might thus be those with a certain amount of both. But it is more likely that phenomena emergent from the interactions must also be considered, as when a single subgroup achieves dominance, the agenda of the group is likely to become less enticing to other subgroups, especially those with higher social distance.

Whether religions can be characterized as complex adaptive systems is still an open question. The case of Finnish revivals demonstrates the emergence of stable systems via processes that can be characterized as self-organizatory. I have also argued that these systems adjusted their behavior in response to environmental pressures, and these reactions were likely to be adaptive in the evolutionary sense. What remains to be shown is that religious systems are also complex, and that there is interdependence between their elements so that removing one will seriously impair the whole system. That group cohesion appears dependent on both deific beliefs and ritual life suggests this, but in the present age there are many perfectly functional groups that feature neither, so arguments on components necessary for cohesion must be formulated with care.

Two strands of research have begun to explore the issue. One, the work of Czachesz (2007, 2012) has cast the connections between religious experiences, beliefs, texts, and rituals in systemic terms. The next step is to demonstrate whether these apparent dependencies are spurious or involve feedback loops or other mechanisms that maintain them despite perturbations. Two, the adaptationist-byproduct debate (see chapter 4) has produced accounts of how religiosity may have evolved as a singular adaptation that integrates several aspects of cognition, behavior, and physiology to provide novel solutions to ecological challenges (Alcorta and Sosis 2005; Purzycki and Sosis 2010).

These questions are difficult to study, because over historical time religious systems have become more diffused with numerous other systems (people who speak the same language do not necessarily share the same religion, and those with the same religion belong to different political systems, etc.), which may have stalled their transition into more integrated systems (Stearns 2007). This makes even demarcating the object of study a difficult task.

7.2.3 Revival Movements' Divergent Trajectories

Regarding the historical study of the revivals, the signaling thesis could be employed to account for the different trajectories of the revival movements that emerged. First, in 1844 the followers of clergyman Fredrik Gabriel Hedberg split from the *Skört* Revival to join the Evangelical movement (*Evankelinen herätysliike*; Haavio 1965, 47–48, 55–56). Later registers of the society reveal that the movement was dominated by merchants and city gentry (Koskenniemi 1967, 62, 110). A prediction from the theory would be that as a movement becomes wider and more diffuse, its members share fewer of the same concerns, making the maintenance of unity more challenging. The split and the differences between the movements could thus be analyzed from this perspective.

Second, a central theoretical idea in the present study is that emotional and ecstatic phenomena work to ratchet up levels of trust. But they are fleeting, and after the initial fervor passes, only costly behavioral signals are employed. The Laestadian Revival appears a curious exception to this rule: within it, the ecstatic trance became institutionalized as an enduring practice of "being moved" (*liikutukset*; Raittila 1976, 193–196). There are however differing reports on how long they actually endured (see Pyysiäinen 2004a, 186), and they may have been on the wane for some decades now (Kulpakko 2007). How such retention is possible is an interesting theoretical and empirical question.

Third, the Laestadian movement appears an exception also in other respects, especially in comparison to the present day *Skörts* or "the Awoken." Whereas in their inception the two were alike in many respects, being cohesive and commanding strict observance of the movement's norms, a survey by Salomäki (2010, 375–378) reveals that while Laestadianism has retained this character, the Awoken movement has become more like a loose network centered around the annual summer festival. In moral attitudes the latter did not appear deviant from the surrounding populace. Other inspected movements fell in between these two. Why have Laestadians and *Skörts* taken such different trajectories over 150 years? A simple answer is suggested by the signaling theory: the *Skört* heartland of Ostrobothnia is one of the more affluent and well-connected regions of Finland, whereas Laestadians have traditionally resided in the north, which is very remote and consequently suffers from poorer services. An illuminating narrative is offered by a Laestadian blog (Torvi 2008): when a car breaks down on a Sunday afternoon, hundreds of kilometers from home, one can only draw on his/her network in getting help. But this is dependent on strong norms of indirect reciprocity. Frequent mutual assistance and strong group norms can thus have upheld one another, whereas within the *Skörts,* availability of other options has given less impetus for maintaining such strong bonds. However, this explanation is subject to the observation that the Laestadian movement is not confined to the desolate north (Salomäki 2010, 204–205).

7.2.4 Revivals and Enclosures

The relationship between environmental changes and religious activity is complex. Antecedent research includes some strong results but does not allow for identification of specific stressors. As of yet these appear mutable and context-dependent. Concerning the history of Finnish revivals, one conspicuously under-researched area is the relationship between the revivals and the enclosure processes. As stated in chapter 5.2.1, the enclosure act appears as a turning point without equal in the economic history of Finland, as it created a new social division between the land owning and the landless classes. It also dismantled extant forms of cooperation within the community while creating new challeges for upholding social cohesion. Thus, theoretically we should expect significant religious activity to follow in its wake.

In general, both progressed through the country in roughly the same order and period (on enclosures see Nohrström 1933, map on p. 108). Prior to 1800, enclosures had been concluded in Northern Savonia, Southern Ostrobothnia, and the Southwestern coast, the three areas with notable early revival activity.[14] By

14 The Savonian *Skört* revival is conventionally seen as beginning in 1796, when haymakers at Telppäs field of Iisalmi parish experienced the “Holy Spirit coming down to them one day so powerfully that they fell down as if dead; and they saw heavenly, fine visions and spoke in different languages, which the Spirit taught them, as in the time of the Apostles. And in the same

1849 only Northern Finland and the Wyborg area were for the most part yet to be parcelled – and the Laestadian Revival didn't spread to these until later in the century. It is also of note that in Tavastia this first phase of enclosures produced fewer changes to the existing order (Saarenheimo 2003, 355–358), and the corresponding lack of (large-scale) revivals in the area is conspicuous. Finally, in some cases the two processes appear to simultaneously occur in a single parish (Toivanen 1997; Sulkunen 1999, 47–53; Kimanen 2008, 185–191).

But seeking a correlation between the two presents considerable difficulties for at least two reasons. First, the enclosure process in a single parish was often protracted, and pinpointing any single year as the postulated "point of rupture" in the fabric of local society may be nigh impossible. Such a research project would involve going through a mass of records on these processes and also require the compilation of an index of the dates of all eighteenth and nineteenth century revivals in the historical record, as none is yet available. Second, the correspondence does not appear completely neat even on the national scale. In the northernmost reaches the enclosures were not finished until the twentieth century, long after a Laestadian Revival had spread to these areas. And in the areas of early enclosures, it is only the much later revivals that appear to have staying-power (the Jumpers and *Skörts*). The latter could have to do with the 1789 act and the reasons stated above (chapter 7.1.1), but at the moment this is highly conjectural.

---

year it spread to other villages little by little, and by December their number was quite big" (Niskanen 1998, 3–4).

8 Sources

Finland's Family History Association (*Suomen Sukuhistoriallinen yhdistys ry.*) digiarchives:

- Kuortane population tables 1774–1798: http://www.digiarkisto.org/sshy/kirjat/Kirkonkirjat/kuortane/vakilukutauluja_1774-1798_va178/hakemisto_1.htm [read 4.8.2011]
- Lappi communion book 1832–1838: http://www.digiarkisto.org/sshy/kirjat/Kirkonkirjat/lappi/rippikirja_1832-1838_mko11-16/hakemisto_1.htm [read 5.11.2010]

The National Archives of Finland (*Kansallisarkisto*) microfilms:

- Kuortane population tables 1802–1847: VÄ177
- Luvia population tables 1805–1877: VÄ113

National Library of Finland (*Kansalliskirjasto*) digital collections:

- Helsingfors Morgonblad 49, June 29th 1835: http://digi.lib.helsinki.fi/sanomalehti/secure/showPage.html?conversationId=6&action=entryPage&id=393667 [read 3.3.2011]
- Helsingfors Morgonblad 50, July 3rd 1835: http://digi.lib.helsinki.fi/sanomalehti/secure/showPage.html?conversationId=3&action=entryPage&id=393720 [read 3.3.2011]
- Helsingfors Morgonblad 51, July 6th 1835: http://digi.lib.helsinki.fi/sanomalehti/secure/showPage.html?conversationId=4&action=entryPage&id=393735 [read 3.3.2011]
- Helsingfors Morgonblad 52, July 10th 1835: http://digi.lib.helsinki.fi/sanomalehti/secure/showPage.html?conversationId=5&action=entryPage&id=393681 [read 3.3.2011]
- Uusi Suometar 234, October 9th 1895: http://digi.lib.helsinki.fi/sanomalehti/secure/showPage.html?action=page&type=lq&conversationId=2&id=593995&pageFrame_currPage=1 [read 12.9.2010]
- Uusi Suometar 235, October 10th 1895: http://digi.lib.helsinki.fi/sanomalehti/secure/showPage.html?action=page&type=lq&conversationId=1&id=593615&pageFrame_currPage=1 [read 12.9.2010]

9 References

Aberle, David. 1965. "A Note on Relative Deprivation Theory As Applied to Millenarian and Other Cult Movements." In *Reader in Comparative Religion: An Anthropological Approach (Second Edition)*, edited by William A. Lessa, and Evon Z. Vogt, 537–541. New York, Evanston & London: Harper & Row.

Abu-Lughod, Janet L. 1989. *Before European Hegemony: The World System A.D. 1250–1350.* Oxford & New York: Oxford University Press.

Ahmed, Ali M. 2009. "Are Religious People More Prosocial? A Quasi-Experimental Study with *Madrasah* Pupils in a Rural Community in India." *Journal for the Scientific Study of Religion* 48:368–374.

Airas, Kaarlo. 1936. "Tornionjokilaakson tärkeimpien hengellisten liikkeitten suhteista 1700-luvulla." In *Suomen kirkkohistoriallisen seuran vuosikirja 1925–1928*, 243–269. Helsinki: Suomen kirkkohistoriallinen seura.

Akiander, Matth. 1859. *Historiska Upplysningar om Religiösa rörelserna i Finland i äldre och senare tider (III. delen).* Helsingfors: Kejserliga Alexanders-Universitetet.

———. 1860. *Historiska Upplysningar om Religiösa rörelserna i Finland i äldre och senare tider (IV. delen).* Helsingfors: Kejserliga Alexanders-Universitetet.

———. 1862. *Historiska Upplysningar om Religiösa rörelserna i Finland i äldre och senare tider (VI. delen).* Helsingfors: Kejserliga Alexanders-Universitetet.

Alcorta, Candace S., and Sosis, Richard. 2005. "Ritual, Emotion, and Sacred Symbols: The Evolution of Religion as an Adaptive Complex." *Human Nature* 16:323–359.

———. 2006. "Why ritual works: A rejection of the by-product hypothesis." *Behavioral and Brain Sciences* 29:613–614.

Allport, Gordon W., and Ross, J. Michael. 1967. "Personal religious orientation and prejudice." *Journal of Personality and Social Psychology* 5:432–443.

Anderson, P. W. 1972. "More Is Different: Broken symmetry and the nature of the hierarchical structure of science." *Science* 177:393–396.

Anttonen, Veikko. 1996. *Ihmisen ja maan rajat. 'Pyhä' kulttuurisena kategoriana.* Helsinki: Suomalaisen kirjallisuuden seura.

Atran, Scott. 2010. *Talking to the Enemy: Faith, Brotherhood, and the (Un)making of Terrorists.* New York: HarperCollins.

Atran, Scott, and Henrich, Joseph. 2010. "The Evolution of Religion: How Cognitive By-Products, Adaptive Learning Heuristics, Ritual Displays, and Group Competition Generate Deep Commitments to Prosocial Religions." *Biological Theory* 5:18–30.

Atran, Scott, and Medin, Douglas. 2008. *The Native Mind and the Cultural Construction of Nature.* Cambridge & London: The MIT Press.

Atran, Scott, and Norenzayan, Ara. 2004. "Religion's evolutionary landscape: Counterintuition, commitment, compassion, communion." *Behavioral and Brain Sciences* 27:713–770.

Axelrod, Robert. [1984] 2006. *The Evolution of Cooperation (Revised Edition).* New York: Basic Books.

Axelrod, Robert, and Hamilton, William D. 1981. "The Evolution of Cooperation." *Science* 211:1390–1396.

Axtell, Robert L. 2000. "Why Agents? On the Varied Motivations for Agent Computing in the Social Sciences." http://www.brookings.edu/es/dynamics/papers/agents/agents.pdf [read 1.4.2010].

Bacharach, Michael, and Gambetta, Diego. 2001. "Trust in Signs." In *Trust in Society*, edited by Karen S. Cook, 148–184. New York: Russell Sage Foundation.

Bainbridge, William Sims. 2006. *God from the Machine: Artificial Intelligence Models of Religious Cognition.* Lanham: Altamira.

Barrett, Justin L. 2004. *Why Would Anyone Believe in God?* Walnut Creek: Altamira.

———. 2007. "Cognitive Science of Religion: What Is It and Why Is It? " *Religion Compass* 1:1–19.

Barrett, Lisa Feldman, Mesquita, Batja, and Gendron, Maria. 2011. "Context in Emotion Perception." *Current Directions in Psychological Science* 20:286–290.

Bateson, Melissa, Nettle, Daniel, and Roberts, Gilbert. 2006. "Cues of being watched enhance cooperation in a real-world setting." *Biology Letters* 2:412–414.

Batson, C. Daniel, Schoenrade, Patricia, and Ventis, W. Larry. 1993. *Religion and the Individual.* New York & Oxford: Oxford University Press.

Bering, Jesse M. 2004. "The Evolutionary History of an Illusion: Religious Causal Beliefs in Children and Adults." In *Origins of the Social Mind: Evolutionary Psychology and Child Development*, edited by Bruce J. Ellis, and David F. Bjorklund, 411–437. New York: Guilford Press.

Bering, Jesse M., McLeod, Katrina, and Shackelford, Todd K. 2005. "Reasoning about Dead Agents Reveals Possible Adaptive Trends." *Human Nature* 16:360–381.

Berman, Eli. 2000. "Sect, Subsidy, and Sacrifice: An Economist's View of Ultra-Orthodox Jews." *The Quarterly Journal of Economics* 65:905–953.

Bicchieri, Cristina. 2006. *The Grammar of Society: The Nature and Dynamics of Social Norms.* Cambridge: Cambridge University Press.

Björklund, Johannes. 1939. *Suomen papisto 1800-luvulla, erityisesti silmälläpitäen pappistarvetta ja sen tyydyttämismahdollisuuksia.* Helsinki: Suomen kirkkohistoriallinen seura.

Bliege Bird, Rebecca, and Smith, Eric Alden. 2005. "Signaling Theory, Strategic Interaction, and Symbolic Capital." *Current Antropology* 46:221–248.

Bliege Bird, Rebecca, Smith, Eric Alden, and Bird, Douglas W. 2001. "The hunting handicap: costly signaling in human foraging strategies." *Behavioral Ecology and Sociobiology* 50:9–19.

Boehm, Christopher. 2000. "Conflict and the Evolution of Social Control." In *Evolutionary Origins of Morality: Cross-Disclipinary Perspectives*, edited by Leonard D. Katz, 79–101. Exeter: Imprint Academic.

Bourdieu, Pierre. 1977. *Outline of a Theory of Practice.* Cambridge: Cambridge University Press.

———. 1997. "The Forms of Capital." In *Education: Culture, Economy and Society*, edited by A. H. Halsey, Hugh Lauder, Philip Brown, and Amy Stuart Wells, 46–58. Oxford: Oxford University Press. Original in John G. Richardson (ed.) 1986, *Handbook of Theory and Research for the Sociology of Education.* Westport: Greenwood Press.

Bowles, Samuel, and Gintis, Herbert. 2003. "Origins of Human Cooperation." In *Genetic and Cultural Evolution of Cooperation*, edited by Peter Hammerstein, 429–443. Cambridge & London: The MIT Press.

Boyarin, Daniel. 1999. *Dying for God: Martyrdom and the Making of Christianity and Judaism.* Stanford: Stanford University Press.

Boyd, Robert, and Richerson, Peter J. 1985. *Culture and the Evolutionary Process.* Chicago: The University of Chicago Press.

———. 1988. "The Evolution of Reciprocity in Sizable Groups." *Journal of Theoretical Biology* 132:337–356.

———. 1992. "Punishment Allows the Evolution of Cooperation (or Anything Else) in Sizable Groups." *Ethology and Sociobiology* 13:171–195.

Boyer, Pascal. 2001. *Religion Explained. The Evolutionary Origins of Religious Thought.* New York: Basic Books.

———. 2006. "Ten Problems in Search of a Research Programme: Towards Integrated Naturalistic Explanations of Human Culture." http://artsci.wustl.edu/~pboyer/PBoyerHomeSite/articles/TenProblems3.1.pdf [read 26.11.2009].

Boyer, Pascal, and Bergstrom, Brian. 2008. "Evolutionary Perspectives on Religion." *Annual Review of Anthropology* 37:111–130.

Boyer, Pascal, and Liénard, Pierre. 2006. "Why ritualized behavior? Precaution systems and action-parsing in developmental, pathological and cultural rituals." *Behavioral and Brain Sciences* 29:595–650 (includes open peer commentary and author response).

Brewer, Marilynn B., and Kramer, Roderick M. 1986. "Choice Behavior in Social Dilemmas: Effects of Social Identity, Group Size, and Decision Framing." *Journal of Personality and Social Psychology* 50:543–549.

Brown, William Michael, Palameta, Boris, and Moore, Chris. 2003. "Are there nonverbal cues to commitment? An exploratory study using the zero-acquaintance video presentation paradigm." *Evolutionary Psychology* 1:42–69.

Bulbulia, Joseph. 2004. "Religious Costs as Adaptations that Signal Altruistic Intention." *Evolution and Cognition* 10:19–42.

———. 2012. "Why "costly-signalling" models of religion require cognitive psychology" (manuscript). In *Origins of Religion, Cognition, and Culture*, edited by A.W. Geertz, and J. Sinding Jensen. London: Equinox. http://www.victoria.ac.nz/religion/downloads/V4Aarhusbook_Bulbulia.pdf [read 10.6.2009].

Bulbulia, Joseph, and Mahoney, Andrew. 2008. "Religious solidarity: The hand grenade experiment." *Journal of Cognition and Culture* 8:295–320.

Bulbulia, Joseph, Sosis, Richard, Harris, Erica, Genet, Russell, Genet, Cheryl, and Wyman, Karen, eds. 2008. *The Evolution of Religion: Studies, Theories, and Critiques.* Santa Margarita: Collins Foundation Press.

Burger, Jerry M. 1999. "The Foot-in-the-Door Compliance Prodecure: A Multiple-Process Analysis and Review." *Personality and Social Psychology Review* 3:303–325.

Burnham, Terence C., and Hare, Brian. 2007. "Engineering Human Cooperation: Does Involuntary Neural Activation Increase Public Goods Contributions? " *Human Nature* 18:88–108.

Caciola, Nancy. 2003. *Discerning Spirits: Divine and Demonic Possession in the Middle Ages.* Ithaca: Cornell University Press.

Carroll, Michael P. 1975. "Revitalization Movements and Social Structure: Some Quantitative Tests." *American Sociological Review* 40:390–401.

Chen, Daniel L. 2010. “Club Goods and Group Identity: Evidence from Islamic Resurgence During the Indonesian Financial Crisis.” *Journal of Political Economy* 118:300–354.

Clauset, Aaron, Shalizi, Cosma Rohilla, and Newman, M. E. J. 2009. “Power-Law Distributions in Empirical Data.” *SIAM Review* 51:661–703.

Cohn, Norman. [1970] 2004. *The Pursuit of the Millennium: Revolutionary Millenarians and Mystical Anarchists of the Middle Ages.* London: Pimlico.

Coleman, James S. 1988. “Social Capital in the Creation of Human Capital.” *The American Journal of Sociology* 94:S95–S120.

Cosmides, Leda, and Tooby, John. 1989. “Evolutionary Psychology and the Generation of Culture, Part II – Case Study: A Computational Theory of Social Exchange.” *Ethology and Sociobiology* 10:51–97.

Cronk, Lee. 1994. “Evolutionary theories of morality and the manipulative use of signals.” *Zygon* 29:81–101.

Czachesz, István. 2007. “The Emergence of Early Christian Religion: A Naturalistic Approach.” In *Explaining Early Judaism and Christianity: Contributions from Cognitive and Social Science*, edited by Petri Luomanen, Ilkka Pyysiäinen, and Risto Uro, 73–94. Leiden & Boston: Brill.

———. 2012. “The Evolutionary Dynamics of Religious Systems: Laying the Foundations of a Network Model” (manuscript). In *Origins of Religion, Cognition and Culture*, edited by A.W. Geertz, and J. Sinding Jensen. London: Equinox. http://religionandcognition.com/publications/czachesz_religious_systems.pdf [read 24.2.2010].

Damasio, Antonio R. 1996. “The Somatic Marker Hypothesis and the Possible Functions of the Prefrontal Cortex.” *Philosophical Transactions: Biological Sciences* 351:1413–1420.

———. 2004. “William James and the Modern Neurobiology of Emotion.” In *Emotion, Evolution, and Rationality*, edited by Dylan Evans, and Pierre Cruse, 3–14. Oxford: Oxford University Press.

Darley, John M., and Batson, C. Daniel. 1973. “‘From Jerusalem to Jericho’: A Study of Situational and Dispositional Variables in Helping Behavior.” *Journal of Personality and Social Psychology* 27:100–108.

Dawes, Robyn M. 1980. “Social Dilemmas.” *Annual Review of Psychology* 31:169–193.

DeJong, William. 1979. “An Examination of Self-Perception Mediation of the Foot-in-the-Door Effect.” *Journal of Personality and Social Psychology* 37:2221–2239.

Dennett, Daniel C. 1995. *Darwin’s Dangerous Idea: Evolution and the Meanings of Life.* London: Penguin Books.

DeScioli, Peter, and Kurzban, Robert. 2009. “Mysteries of morality.” *Cognition* 112:281–299.

Donner, Fred M. 2010. *Muhammad and the Believers: At the Origins of Islam.* Cambridge & London: The Belknap Press of Harvard University Press.

Douglas, Mary. 1986. *How Institutions Think.* Syracuse: Syracuse University Press.

Drea, Christine, and Carter, Allisa N. 2009. “Cooperative problem solving in a social carnivore.” *Animal Behaviour* 78:967–977.

Dunderberg, Ismo. 2008. *Beyond Gnosticism: Myth, Lifestyle, and Society in the School of Valentinus.* New York: Columbia University Press.

Durkheim, Émile. [1895] 1982. *The Rules of Sociological Method (translated by W. D. Halls).* New York: Free Press.

———. 1915. *The Elementary Forms of the Religious Life.* London: George Allen & Unwin Ltd.

Dutton, Donald G., and Aron, Arthur P. 1974. "Some evidence for heightened sexual attraction under conditions of high anxiety." *Journal of Personality and Social Psychology* 30:510 – 517.

Edgell, Penny, Gerteis, Joseph, and Hartmann, Douglas. 2006. "Atheists as 'Other': Moral Boundaries and Cultural Membership in American Society." *American Sociological Review* 71:211 – 234.

Ekman, Paul. 1994. "Strong Evidence for Universals in Facial Expressions: A Reply to Russel's Mistaken Critique." *Psychological Bulletin* 115:268 – 287.

———. 1999. "Basic Emotions." In *Handbook of Cognition and Emotion*, edited by Tim Dalgleish, and Mick Power. Sussex: John Wiley & Sons, Ltd.

Ekman, Paul, Friesen, Wallace V., and O'Sullivan, Maureen. 1988. "Smiles When Lying." *Journal of Personality and Social Psychology* 54:414 – 420.

Ekman, Paul, O'Sullivan, Maureen, and Frank, Mark G. 1999. "A Few Can Catch a Liar." *Psychological Science* 10:263 – 266.

Ellickson, Robert C. 1991. *Order without Law: How Neighbors Settle Disputes.* Cambridge & London: Harvard University Press.

Emmons, Robert A., and McNamara, Patrick. 2006. "Sacred Emotions and Affective Neuroscience: Gratitude, Costly Signaling, and the Brain." In *Where God and Science Meet – How Brain and Evolutionary Studies Alter Our Understanding of Religion, Volume 1: Evolution, Genes, and the Religious Brain*, edited by Patrick McNamara, 11 – 30. Westport: Praeger Publishers.

Engelmann, Dirk, and Fischbacher, Urs. 2002. "Indirect Reciprocity and Strategic Reputation Building in an Experimental Helping Game." http://www.iew.uzh.ch/wp/iewwp132.pdf [read 19. 5. 2011]. University of Zurich, Institute for Empirical Research in Economics, Working paper no. 132.

Ensminger, Jean. 1997. "Transaction Costs and Islam: Conversion in Africa." *Journal of Institutional and Theoretical Economics* 153:4 – 29.

Epstein, Joshua M., and Axtell, Robert. 1996. *Growing Artificial Societies: Social Science from the Bottom Up.* Washington: The Brookings Institution.

Ericsson, Tom, and Harnesk, Börje. 1994. *Präster, predikare och profeter: Läseriet i övre Norrland.* Gideå: Vildros.

Fehr, Ernst, and Fischbacher, Urs. 2005. "The Economics of Strong Reciprocity." In *Moral Sentiments and Material Interests: The Foundations of Cooperation in Economic Life*, edited by Herbert Gintis, Samuel Bowles, Robert Boyd, and Ernst Fehr, 151 – 191. Cambridge: MIT Press.

Fehr, Ernst, and Gächter, Simon. 2002. "Altruistic punishment in humans." *Nature* 415:137 – 140.

Fessler, Daniel M. T., and Haley, Kevin J. 2003. "The Strategy of Affect: Emotions in Human Cooperation." In *Genetic and Cultural Evolution of Cooperation*, edited by Peter Hammerstein, 7 – 36. Cambridge & London: The MIT Press.

Fishman, Aryei, and Goldschmidt, Yaaqov. 1990. "The Orthodox Kibbutzim and Economic Success." *Journal for the Scientific Study of Religion* 29:505 – 511.

Frank, Robert H. 1993. "Strategic Role of the Emotions; Reconciling Over- and Undersocialized Accounts of Behavior." *Rationality and Society* 5:160 – 184.

Freedman, Jonathan L., and Fraser, Scott C. 1966. "Compliance without Pressure: The Foot-in-the-Door Technique." *Journal of Personality and Social Psychology* 4:195 – 202.

French, Steven, and Ladyman, James. 1999. "Reinflating the Semantic Approach." *International Studies in the Philosophy of Science* 13:103–121.
Frith, Chris, and Frith, Uta. 2005. "Theory of Mind." *Current Biology* 15:644–645.
Fräntilä, Markku. 2001. "Yhtenäiskulttuurin uhri: Pohjanmaan hurmahenkinen separatisti Jaakko Wallenberg." In *Perinnettä vai bisnestä? Kulttuurin paikalliset ulottuvuudet*, edited by Sulevi Riukulehto, 179–202. Jyväskylä: Atena Kustannus Oy.
Gavrilets, Sergey, Anderson, David G., and Turchin, Peter. 2010. "Cycling in the Complexity of Early Societies." *Cliodynamics* 1.
Gervais, Will M., and Henrich, Joseph. 2010. "The Zeus Problem: Why Representational Content Biases Cannot Explain Faith in Gods." *Journal of Cognition and Culture* 10:383–389.
Gil-White, Francisco J. 2001. "Are Ethnic Groups Biological 'Species' to the Human Brain?" *Current Anthropology* 42:515–554.
Gilbert, Daniel T., and Malone, Patrick S. 1995. "The Correspondence Bias." *Psychological Bulletin* 117:21–38.
Gilbert, Nigel. 1995. "Emergence in Social Simulation." In *Artificial Societies: The Computer Simulation of Social Life*, edited by Nigel Gilbert, and Rosaria Conte, 122–131. London: University College London Press.
Ginges, Jeremy, Hansen, Ian, and Norenzayan, Ara. 2009. "Religion and Support for Suicide Attacks." *Psychological Science* 20:224–230.
Gintis, Herbert. 2004. "Towards the Unity of the Behavioral Sciences." *Politics, Philosophy and Economics* 3:37–57.
Gintis, Herbert, Smith, Eric Alden, and Bowles, Samuel. 2001. "Costly Signaling and Cooperation." *Journal of Theoretical Biology* 213:103–119.
Glock, Charles Y. 1964. "The Role of Deprivation in the Origin and Evolution of Religious Groups." In *Religion and Social Conflict: Based upon Lectures given at the Institute of Ethics and Society at San Francisco Theological Seminary*, edited by Robert Lee, and Martin E. Marty, 24–36. New York: Oxford University Press.
Glock, Charles Y., and Stark, Rodney. 1965. *Religion and Society in Tension.* Chicago: Rand McNally & Company.
Granovetter, Mark S. 1973. "The Strength of Weak Ties." *The American Journal of Sociology* 78:1360–1380.
Gray, Kurt, and Wegner, Daniel. 2010. "Blaming God for Our Pain: Human Suffering and the Divine Mind." *Personality and Social Psychology Review* 14:7–16.
Greif, Avner. 1989. "Reputation and Coalitions in Medieval Trade: Evidence on the Maghribi Traders." *The Journal of Economic History* 49:857–882.
———. 1993. "Contract Enforceability and Economic Institutions in Early Trade: The Maghribi Traders' Coalition." *The American Economic Review* 83:525–548.
Greus, Aija, and Hirvelä, Harri. 2006. "Historismi.net: Ammatit." http://www.saunalahti.fi/hirvela/historismi_sivut/ammatit a-k.html [read 16.5.2011].
Griffits, Paul E. 2004. "Towards a 'Machiavellian' Theory of Emotional Appraisal." In *Emotion, Evolution, and Rationality*, edited by Dylan Evans, and Pierre Cruse, 89–105. Oxford: Oxford University Press.
Guthrie, Stewart Elliott. 1993. *Faces in the Clouds: A New Theory of Religion.* New York & Oxford: Oxford University Press.
Gächter, Simon. 2006. "Conditional cooperation: Behavioral regularities from the lab and the field and their policy implications." Paper provided by The Centre for Decision Research

and Experimental Economics, School of Economics, University of Nottingham in its series Discussion Papers with number 2006–03. http://ideas.repec.org/p/cdx/dpaper/2006-03.html [read 19.5.2011].

Haatanen, Pekka. 1968. *Suomen maalaisköyhälistö tutkimusten ja kaunokirjallisuuden valossa.* Helsinki: Werner Söderström Osakeyhtiö.

Haavio, Ari. 1965. *Suomen uskonnolliset liikkeet.* Porvoo & Helsinki: Werner Söderström Osakeyhtiö.

Haidt, Jonathan. 2003. "The moral emotions." In *Handbook of Affective Sciences*, edited by R. Davidson, K. Sherer, and H. H. Goldsmith, 852–870. New York: Oxford University Press.

Halevy, Nir, Chou, Eileen Y., Cohen, Taya R., and Livingston, Robert W. 2011. "Status Conferral in Intergroup Social Dilemmas: Behavioral Antecedents and Consequences of Prestige and Dominance." *Journal of Personality and Social Psychology* (advance online publication, doi: 10.1037/a0025515).

Haley, Kevin J., and Fessler, Daniel M. T. 2005. "Nobody's watching? Subtle cues cues affect generosity in an anonymous economic game." *Evolution and Human Behavior* 26:245–256.

Hamann, Stephan. 2012. "Mapping discrete and dimensional emotions onto the brain: controversies and consensus." *Trends in Cognitive Sciences* 16:458–466.

Hammerstein, Peter, ed. 2003. *Genetic and Cultural Evolution of Cooperation.* Cambridge & London: The MIT Press.

Hardin, Garrett. 1968. "The Tragedy of the Commons." *Science* 13:1243–1248.

Harris, Marvin. 1978. *Cows, Pigs, Wars & Witches: The Riddles of Culture.* Glasgow: Fontana Books.

Hartikainen, Esko. 2005. *Heränneitä ja nukahtaneita: Kulttuuri, kontrolli ja herätys 1800-luvun alun Liperissä.* Helsinki: Suomalaisen kirjallisuuden seura.

Haselton, Martie G., and Nettle, Daniel. 2006. "The Paranoid Optimist: An Integrative Evolutionary Model of Cognitive Biases." *Personality and Social Psychology Review* 10:47–66.

Hayden, Brian. 1987. "Alliances and Ritual Ecstasy: Human Responses to Resource Stress." *Journal for the Scientific Study of Religion* 26:81–91.

Heikkilä, Markku, and Seppo, Juha. 1987. "Uskonnollinen liike, esivalta ja "maailma"." In *Kansa liikkeessä*, edited by Risto Alapuro, Ilkka Liikanen, Kerstin Smeds, and Henrik Stenius, 70–90. Vaasa: Kirjayhtymä.

Heino, Harri. 1976. *Hyppyherätys – Länsi-Suomen rukoilevaisuuden synnyttäjä.* Helsinki: Suomen kirkkohistoriallinen seura.

———. 1977. "Näkökohtia herätysliikkeen tutkimusmetodin määräytymisestä." *Teologinen aikakauskirja* 34–40.

Helander, Eila. 1987. *Naiset eivät vaienneet: Naisevankelistainstituutio Suomen helluntailiikkeessä.* Helsinki: Suomen kirkkohistoriallinen seura.

Henrich, Joseph. 2004. "Cultural group selection, coevolutionary processes and large-scale cooperation." *Journal of Economic Behavior & Organization* 53:3–35.

———. 2009. "The evolution of costly displays, cooperation and religion: credibility enhancing displays and their implications for cultural evolution." *Evolution and Human Behavior* 30:244–260.

Henrich, Joseph, and Boyd, Robert. 2001. "Why People Punish Defectors: Weak Conformist Transmission can Stabilize Costly Enforcement of Norms in Cooperative Dilemmas." *Journal of Theoretical Biology* 208:79–89.

Henrich, Joseph, Boyd, Robert, Bowles, Samuel, Camerer, Colin, Fehr, Ernst, and Gintis, Herbert, eds. 2004. *Foundations of Human Sociality: Economic Experiments and Ethnographic Evidence from Fifteen Small-Scale Societies.* Oxford: Oxford University Press.

Henrich, Joseph, Boyd, Robert, Bowles, Samuel, Camerer, Colin, Fehr, Ernst, Gintis, Herbert, and McElreath, Richard. 2001. "In Search of Homo Economicus: Behavioral Experiments in 15 Small-Scale Societies." *The American Economic Review* 91:73–78.

Henrich, Joseph, Ensminger, Jean, McElreath, Richard, Barr, Abigail, Barrett, Clark, Bolyanatz, Alexander, Cardenas, Juan Camilo, Gurven, Michael, Gwako, Edwins, Henrich, Natalie, Lesorogol, Carolyn, Marlowe, Frank, Tracer, David, and Ziker, John. 2010. "Markets, Religion, Community Size, and the Evolution of Fairness and Punishment." *Science* 327:1480–1484.

Henrich, Joseph, Heine, Steven J., and Norenzayan, Ara. 2010. "The Weirdest People in the World? " *Behavioral and Brain Sciences* 33:1–75.

Henrich, Joseph, and McElreath, Richard. 2003. "The Evolution of Cultural Evolution." *Evolutionary Anthropology* 12:123–135.

Henrich, Joseph, McElreath, Richard, Barr, Abigail, Ensminger, Jean, Barrett, Clark, Bolyanatz, Alexander, Cardenas, Juan Camilo, Gurven, Michael, Gwako, Edwins, Henrich, Natalie, Lesorogol, Carolyn, Marlowe, Frank, Tracer, David, and Ziker, John. 2006. "Costly Punishment Across Human Societies." *Science* 312:1767–1770.

Henrich, Natalie, and Henrich, Joseph. 2007. *Why Humans Cooperate: A Cultural and Evolutionary Explanation.* Oxford: Oxford University Press.

Herrmann, Benedikt, Thöni, Christian, and Gächter, Simon. 2008. "Antisocial Punishment Across Societies." *Science* 319:1362–1367.

Hietaniemi, Tapani. 2005. "Ihmistoiminnan verkon arkkitehdit (esipuhe)." J. R. McNeill & William H. McNeill, *Verkottunut ihmiskunta: Yleiskatsaus maailmanhistoriaan*, 13–15. Tampere: Vastapaino.

Hirschfeld, Lawrence A., and Gelman, Susan A., eds. 1994. *Mapping the Mind: Domain Specificity in Cognition and Culture.* Cambridge: Cambridge University Press.

Hirvelä, Harri. 2009. "Historismi.net: Mitat ja rahat." http://www.saunalahti.fi/hirvela/historismi_sivut/mitatsivu2.html [read 14.10.2010].

Hirvonen, Pekka. 2000. "Suku, uskonto, talous, politiikka – eliittien verkostoja 1800-luvun Suomessa." *Historiallisia Papereita* 6. http://www.helsinki.fi/historia/yhdistys/vanhat/julk/verkko00/hirvonen.html [read 3.6.2010].

———. 2003. *Papin pojasta suurteollisuusmieheksi: Alfred Kihlman ja nousevan keskiluokan sosiaaliset verkostot 1800-luvun Suomessa.* Unpublished licentiate thesis. http://urn.fi/URN:NBN:fi-fe20031192 [read 11.2.2009].

Hoge, Dean R., and Roozen, David A. 1979. "Research on Factors Influencing Church Commitment." In *Understanding Church Growth and Decline 1950–1978*, edited by Dean R. Hoge, and David A. Roozen, 42–68. New York: The Pilgrim Press.

Holden, Andrew. 2002a. "Averting Risk: A Cultural Analysis of the Worldview of Jehovah's Witnesses." Department of Sociology, Lancaster University, Lancaster LA1 4YN.

http://www.lancs.ac.uk/fass/sociology/papers/holden-averting-risk.pdf [read 23.3.2009].
———. 2002b. *Jehovah's Witnesses: Portrait of a Contemporary Religious Movement.* London: Routledge.
Hooper, Paul L., Kaplan, Hillard S., and Boone, James L. 2010. "A theory of leadership in human cooperative groups." *Journal of Theoretical Biology* 265:633–646.
Hopkirk, Peter. [1982] 1995. *Trespassers on the Roof of the World: The Secret Exploration of Tibet.* New York, Tokyo & London: Kodansha International.
Houser, Daniel, McCabe, Kevin, and Smith, Vernon. 2004. "Discussion: Cultural group selection, co evolutionary processes and large-scale cooperation (by Joseph Henrich)." *Journal of Economic Behavior & Organization* 53:85–88.
Howell, David C. 2002. *Statistical Methods for Psychology (Fifth edition).* Pacific Grove: Duxbury.
Huhta, Ilkka. 2007. *"Täällä on oikea Suomenkansa": Körttiläisyyden julkisuuskuva 1880–1918.* Helsinki: Suomen kirkkohistoriallinen seura.
Humphreys, Paul. 2009. "The Philosophical Novelty of Computer Simulation Methods." *Synthese* 169:615–626.
Hunsberger, Bruce, and Ennis, Jeff. 1982. "Experimenter Effects in Studies of Religious Attitudes." *Journal for the Scientific Study of Religion* 21:131–137.
Häkkinen, Antti, Ikonen, Vappu, Pitkänen, Kari, and Soikkanen, Hannu. 1991. *Kun halla nälän tuskan toi: miten suomalaiset kokivat 1860-luvun nälkävuodet.* Porvoo: WSOY.
Hökkä, Tuula. 2006. *Margareta Högman: Ystäväliike ja arkkivirsi.* Helsinki: Studio Arkki.
Iannaccone, Laurence R. 1992. "Sacrifice and Stigma: Reducing Free-riding in Cults, Communes, and Other Collectives." *Journal of Political Economy* 100:271–291.
———. 1994. "Why Strict Churches Are Strong." *The American Journal of Sociology* 99:1180–1211.
Ibn Khaldûn, 'Abd-ar-Rahmân Abû Zayd bin Muhammad bin Muhammad. [1377] 1967. *The Muqaddimah: An Introduction to History (translated by Franz Rosenthal, edited and abridged by N. J. Dawood).* Princeton: Princeton University Press.
Irons, William. 1996. "In Our Own Self-image: The Evolution of Morality, Deception, and Religion." *Skeptic* 4:50–61.
———. 2001. "Religion as a Hard-to-Fake Sign of Commitment." In *Evolution and the Capacity for Commitment*, edited by Randolph M. Nesse, 292–309. New York: Russel Sage Foundation.
Isichei, Elizabeth Allo. 1964. "From Sect to Denomination in English Quakerism, with Special Reference to the Nineteenth Century." *The British Journal of Sociology* 15:207–222.
———. 1970. *Victorian Quakers.* London: Oxford University Press.
Iyer, Gopalkrishnan R. 1999. "The impact of religion and reputation in the organization of Indian merchant communities." *Journal of Business & Industrial Marketing* 14:102–117.
Jenkins, Philip. 2008. *The Lost History of Christianity: The Thousand-Year Golden Age of the Church in the Middle East, Africa, and Asia.* Oxford: Lion Hudson.
Jin, Emily M., Girvan, Michelle, and Newman, M. E. J. 2001. "Structure of Growing Social Networks." *Physical Review E* 64:1–8.
Johnson, Dominic, and Bering, Jesse. 2009. "Hand of God, Mind of Man: Punishment and Cognition in the Evolution of Cooperation." In *The Believing Primate: Scientific, Philosophical, and Theological Reflections on the Origin of Religion*, edited by Jeffrey Schloss, and Michael Murray, 26–43. Oxford: Oxford University Press.

Johnson, Dominic D. P. 2005. "God's Punishment and Public Goods: A Test of the Supernatural Punishment Hypothesis in 186 World Cultures." *Human Nature* 16:410–446.

Johnson, Dominic D. P., and Krüger, Oliver. 2004. "The Good of Wrath: Supernatural Punishment and the Evolution of Cooperation." *Political Theology* 5:159–176.

Johnson, Dominic P. 2011. "Why God is the best punisher." *Religion, Brain & Behavior* 1:77–84.

Jutikkala, Eino. 1949. "Varallisuussuhteet Suomessa Ruotsinajan päättyessä." *Historiallinen aikakauskirja* 170–206.

———. 1958. *Suomen talonpojan historia.* Helsinki: Suomalaisen kirjallisuuden seura.

Juva, Mikko. 1962. "Protestiliikkeet Suomen kirkkohistoriassa." *Teologinen aikakauskirja* 185–195.

Kallgren, Carl A., Reno, Raymond R., and Cialdini, Robert B. 2000. "A Focus Theory of Normative Conduct: When Norms Do and Do not Affect Behavior." *Personality and Social Psychology Bulletin* 26:1002–1012.

Kansanen, Toivo. 1915. "'Vasulaisuudesta' Kuortaneella 1800-luvulla." *Vartija* 149–154.

Kanter, Rosabeth Moss. 1972. *Commitment and Community: Communes and Utopias in Sociological Perspective.* Cambridge & London: Harvard University Press.

Karisto, Antti, Takala, Pentti, and Haapola, Ilkka. 1997. *Matkalla nykyaikaan: Elintason, elämäntavan ja sosiaalipolitiikan muutos Suomessa.* Helsinki: WSOY.

Kauranen, Kaisa. 1999. *Rahvas, kauppias ja virkamiehet: Katovuodet Pohjois-Suomessa 1830-luvulla.* Helsinki: Helsingin yliopisto, Yhteiskuntahistorian laitos. Julkaisematon lisensiaatintutkielma.

Kearns, Michael, Judd, Stephen, Tan, Jinson, and Wortman, Jennifer. 2009. "Behavioral experiments on biased voting in networks." *Proceedings of the National Academy of Sciences* 106:1347–1352.

Keinan, Giora. 1994. "Effects of Stress and Tolerance of Ambiguity on Magical Thinking." *Journal of Personality and Social Psychology* 67:48–55.

Kimanen, Anuleena. 2008. *Herätys kylässä ja tuvissa: Rääkkylän Oravisalon kylä ja renqvistiläinen herätys 1820-luvulla.* Helsinki: Suomen kirkkohistoriallinen seura.

King, Andrew J., Johnson, Dominic D. P., and van Vugt, Mark. 2009. "The Origins and Evolution of Leadership." *Current Biology* 19:911–916.

Kinnunen, Mauri. 2004. *Herätysliike kahden kulttuurin rajalla: Lestadiolaisuus Karjalassa 1870–1939.* Jyväskylä: Jyväskylän yliopisto.

Kirby, Maurice. 1993. "Quakerism, entrepreneurship and the family firm in North-East England." In *Entrepreneurship, networks and modern business*, edited by Jonathan Brown, and Mary B. Rose, 105–126. Manchester: Manchester University Press.

Kitcher, Philip, and Vickers, A. Leah. 2003. "Pop Sociobiology Reborn: The Evolutionary Psychology of Sex and Violence." In *Evolution, Gender, and Rape*, edited by Cheryl Brown Travis, 139–168. Cambridge: The MIT Press.

Klemetti, Heikki. 1932. *Kuortaneen vaiheita sanoin ja kuvin: Muistojulkaisu pitäjän 300-vuotisjuhlaan 1932.* Helsinki: Werner Söderström Osakeyhtiö.

Knuuttila, Tarja. 2005. "Models, Representation, and Mediation." *Philosophy of Science* 72:1260–1271.

Koitti, Paavo. 1976. "Margareta Högman, hänen ystävänsä ja vastustajansa." In *Joutsan kirja*, edited by Matti Musikka, 535–552. Joutsa: Joutsan kunta ja seurakunta.

Kokko, Hanna. 2007. *Modelling for Field Biologists and Other Interesting People*. Cambridge: Cambridge University Press.

Korhonen, Teppo. 2003. "Koneiden aika alkamassa." In *Suomen maatalouden historia – Osa 1: Perinteisen maatalouden aika, esihistoriasta 1870-luvulle*, edited by Viljo Rasila, Eino Jutikkala, and Anneli Mäkelä-Alitalo, 461–. Helsinki: Suomalaisen kirjallisuuden seura.

Koskenniemi, Lauri. 1967. *Suomen evankelinen liike 1870–1895*. Helsinki: Suomen luterilainen evankeliumiyhdistys.

Koskinen, Seppo, Martelin, Tuija, Notkola, Irma-Leena, Notkola, Veijo, Pitkänen, Kari, Jalovaara, Marika, Mäenpää, Elina, Ruokolainen, Anne, Ryynänen, Markku, and Söderling, Ismo, eds. 2007. *Suomen väestö*. Helsinki: Gaudeamus.

Krook, Tor, and Salomies, Ilmari, eds. 1937. *Suomen uskonnollisten liikkeiden historiasta: asiakirjoja ja tutkimuksia (toinen nidos)*. Helsinki: Suomen kirkkohistoriallinen seura.

Kuisma, Markku. 2006. *Metsäteollisuuden maa: Suomi, metsät ja kansainvälinen järjestelmä 1620–1920*. Helsinki: Suomalaisen kirjallisuuden seura.

Kulpakko, Toivo. [1992] 2007. "Nimetön." http://p2.foorumi.info/keskusteluavanhoillislestadiolaisuudesta/viewtopic.php?t=314 [read 10.8.2011]. Originally published in *Lapin kansa* 28.12.1992.

Kuusi, Sakari. 1933. "Isojako Suomessa vuosina 1757–1809." In *Suomen maanmittauksen historia I: Ruotsinvallan aika*. Porvoo: Werner Södeström osakeyhtiö.

Kyntäjä, Timo. 2000. "The Lutheran Ascetism – Does Weber tally with Finland?" In *Jäljillä: Kirjoituksia historian ongelmista, osa 2 (Juhlakirja Jorma Kalelalle hänen 60-vuotispäivänään 12.11.2000*, edited by Pauli Kettunen, Auli Kultanen, and Timo Soikkanen, 195–220. Turku: Kirja-Aurora / Turun yliopisto.

Kääriäinen, Kimmo, Niemelä, Kati, and Ketola, Kimmo, eds. 2003. *Moderni kirkkokansa: Suomalaisten uskonnollisuus uudella vuosituhannella*. Tampere: Kirkon tutkimuskeskus.

Lachmann, Michael, Számadó, Szabolcs, and Bergstrom, Carl T. 2001. "Cost and conflict in animal signals and human language." *Proceedings of the National Academy Sciences of the United States of America* 98:13189–13194.

Laland, Kevin N., Odling-Smee, John, and Feldman, Marcus W. 2000. "Niche construction, biological evolution, and cultural change." *Behavioral and Brain Sciences* 23:131–175.

Lambert, Nathaniel M., Fincham, Frank D., Stillman, Tyler F., Graham, Steven M., and Beach, Steven R.H. 2010. "Motivating Change in Relationships: Can Prayer Increase Forgiveness?" *Psychological Science* 21:126–132.

Lang, Helen S. 1992. *Aristotle's Physics and Its Medieval Varieties*. Albany: State University of New York.

Latané, Bibb, and Darley, John M. 1968. "Group Inhibition of Bystander Intervention in Emergencies." *Journal of Personality and Social Psychology* 10:215–221.

Lawrence, C. H. 1989. *Medieval Monasticism: Forms of Religious Life in Western Europe in the Middle Ages (2nd edition)*. London & New York: Longman.

Lawson, E. Thomas, and McCauley, Robert N. 1990. *Rethinking Religion: Connecting Cognition and Culture*. Cambridge & New York: Cambridge University Press.

Leach, Edmund. 1954. *Political Systems of Highland Burma: A Study of Kachin Social Structure*. Boston: Beacon Press.

Leoncini, Isabelle, Le Conte, Yves, Costagliola, Guy, Plettner, Erika, Toth, Amy L., Wang, Mianwei, Huang, Zachary, Becard, Jean-Marc, Crauser, Didier, Slessor, Keith N., and Robinson, Gene E. 2004. "Regulation of behavioral maturation by a primer pheromone produced by adult worker honey bees." *Proceedings of the National Academy of Sciences of the United States of America* 101:17559–17564.

Lernia, Savino di. 2006. "Building monuments, creating identity: Cattle cult as a social response to rapid environmental changes in the Holocene Sahara." *Quaternary International* 151:50–62.

Lessnoff, Michael H. 1994. *The Spirit of Capitalism and the Protestant Ethic: An Enquiry into the Weber Thesis.* Aldershot & Brookfield: Edward Elgar.

Liddell, Henry George, and Scott, Robert. 1940. *A Greek-English Lexicon, revised and augmented throughout by Sir Henry Stuart Jones with the assistance of Roderick McKenzie.* Oxford: Clarendon Press. Accessed via Perseus Digital Library, http://www.perseus.tufts.edu/hopper/ [read 26.10.2011].

Lindquist, Kristen A., Wager, Tor D., Kober, Hedy, Bliss-Moreau, Eliza, and Barrett, Lisa Feldman. 2012. "The brain basis of emotion: A meta-analytic review." *Behavioral and Brain Sciences* 35:121–202 (includes open peer commentary and authors' response).

Liénard, Pierre, and Boyer, Pascal. 2006. "Whence collective rituals? A cultural selection model of ritualized behavior." *American Anthropologist* 108:814–827.

Lofland, John, and Stark, Rodney. 1965. "Becoming a world-saver: A theory of conversion to a deviant perspective." *American Sociological Review* 30:862–875.

Lohi, Seppo. 1997. *Pohjolan kristillisyys: Lestadiolaisuuden leviäminen Suomessa 1870–1899.* Oulu: Suomen Rauhanyhdistysten Keskusyhdistys ry.

Lugo, Luis, Stencel, Sandra, Green, John, Shah, Timothy S., Grim, Brian J., Smith, Gregory, Ruby, Robert, Pond, Allison, Kohut, Andrew, Taylor, Paul, and Keeter, Scott. 2006. *Spirit and Power: A 10-Country Survey of Pentecostals.* Washington: The Pew Forum on Religion & Public Life. http://pewforum.org/surveys/pentecostal/ [read 9.6.2009].

Luhmann, Niklas. 2006. "System as Difference." *Organization* 13:37–57.

Luukkanen, Tarja-Liisa. 2001. "Näkymättömän kulttuurin vaikuttavuus: Pohjanmaan mystiikka sosiaalisten innovaatioiden kasvualustana." In *Perinnettä vai bisnestä? Kulttuurin paikalliset ulottuvuudet*, edited by Sulevi Riukulehto, 169–178. Jyväskylä: Atena Kustannus Oy.

MacCulloch, Diarmaid. 2003. *Reformation: Europe's House Divided 1490–1700.* London: Allen Lane & Penguin Group.

Macy, Michael W., and Skvoretz, John. 1998. "The Evolution of Trust and Cooperation between Strangers: A Computational Model." *American Sociological Review* 63:638–660.

Mahoney, Andrew. 2008. "Theological Expressions as Costly Signals of Religious Commitment." In *The Evolution of Religion: Studies, Theories, and Critiques*, edited by Joseph Bulbulia, Richard Sosis, Erica Harris, Russell Genet, Cheryl Genet, and Karen Wyman, 161–166. Santa Margarita: Collins Foundation Press.

Malley, Brian, and Barrett, Justin. 2003. "Can ritual form be predicted from religious belief? A test of the Lawson-McCauley hypotheses." *Journal of Ritual Studies* 17:1–14.

Malley, Brian E. 1995. "Explaining order in religious systems." *Method & Theory in the Study of Religion* 7:5–22.

———. 1997. "Causal holism in the evolution of religious ideas: A reply to Pascal Boyer." *Method & Theory in the Study of Religion* 9:389–399.

Malmivaara, Wilh. 1914. *Puoli Vuosisataa Heränneiden Keskuudessa: Muistelmia.* Tampere: Isak Julinin Kirjapaino.

Marcus, Joyce, and Flannery, Kent V. 2004. "The Coevolution of Ritual and Society: New 14C Dates from Ancient Mexico." *Proceedings of the National Academy of Sciences of the United States of America* 101:18257–18261.

Markkanen, Erkki. 1977. *Maaseutuväestön varallisuusolot ja luottosuhteet Sisä-Suomessa elinkeinoelämän murroskaudella v. 1850–1914.* Jyväskylä: Jyväskylän yliopisto.

Marx, Karl. [1843] 1977. "A Contribution to the Critique of Hegel's 'Philosophy of Right': Introduction." In *Critique of Hegel's 'Philosophy of Right'*, edited by Joseph O'Malley, 129–142. Cambridge, New York & Melbourne: Cambridge University Press.

McCann, Stewart J. H. 1999. "Threatening Times and Fluctuations in American Church Memberships." *Personality and Social Psychology Bulletin* 25:325–336.

McCauley, Robert N., and Lawson, E. Thomas. 2002. *Bringing Ritual to Mind: Psychological Foundations of Cultural Forms.* Cambridge: Cambridge University Press.

McElreath, Richard, and Boyd, Robert. 2007. *Mathematical Models of Social Evolution: A Guide for the Perplexed.* Chicago & London: The University of Chicago Press.

McIntyre, Lee. 2007. "Emergence and reduction in chemistry: ontological or epistemological concepts?" *Synthese* 155:337–343.

McLoughlin, William G. 1978. *Revivals, Awakenings, and Reform: An Essay on Religion and Social Change in America, 1607–1977.* Chicago & London: The University of Chicago Press.

McNeill, J. R., and McNeill, William H. 2003. *The Human Web: A Bird's-eye View of World History.* New York & London: W. W. Norton & Co.

McNeill, William Hardy. [1976] 1998. *Plagues and Peoples.* New York: Anchor Books, Doubleday.

Medin, Douglas L., and Atran, Scott, eds. 1999. *Folkbiology.* Cambridge: MIT Press.

Michod, Richard E. 2007. "Evolution of individuality during the transition from unicellular to multicellular life." *Proceedings of the National Academy of Sciences* 104:8613–8618.

Miller, John H., and Page, Scott E. 2007. *Complex Adaptive Systems: An Introduction to Computational Models of Social Life.* Princeton & Oxford: Princeton University Press.

Monsma, Stephen V. 2007. "Religion and Philanthropic Giving and Volunteering: Building Blocks for Civic Responsibility." *Interdisciplinary Journal of Research on Religion* 3:1–28.

Murray, Michael J., and Moore, Lyn. 2009. "Costly Signaling and the Origin of Religion." *Journal of Cognition and Culture* 9:225–245.

Murtorinne, Eino. 1986. *Suomalainen teologia autonomian kautena (1828–1918).* Helsinki: Gaudeamus.

Myllyntaus, Timo, and Åström, Sven-Erik. 1980. "Suomen talouspolitiikka ja valtiontalous 1809–1860." In *Suomen taloushistoria 1: Agraarinen Suomi*, edited by Eino Jutikkala, Yrjö Kaukiainen, and Sven-Erik Åström, 333–366. Helsinki: Tammi.

Mäki, Uskali. 2009. "Models and truth: The functional decomposition approach." http://www.helsinki.fi/tint/maki/materials/ModelsEPSApubl09.pdf [read 11.3.2011].

Nelson, Robert H. 2010. "Max Weber Revisited." In *Religion, Economy, and Cooperation*, edited by Ilkka Pyysiäinen, 157–218. Berlin & New York: De Gruyter.

Nesse, Randolph M., ed. 2001. *Evolution and the Capacity for Commitment.* New York: Russel Sage Foundation.

Nickerson, Raymond S. 1998. "Confirmation Bias: A Ubiquitous Phenomenon in Many Guises." *Review of General Psychology* 2:175–220.

Niebuhr, H. Richard. [1929] 1987. *The Social Sources of Denominationalism.* Gloucester: Peter Smith.
Niemelä, Kati, and Salomäki, Hanna. 2006. "Herätysliikkeet 2000-luvun kirkollisessa kentässä." *Teologinen aikakauskirja* 111:359 – 371.
Niskanen, Lars Johan. [1837] 1998. *Muisto Kirja Hengellisten asiain päälle.* A facsimile edition published by Matti Pehkonen.
Nohrström, Walter. 1933. "Isojaot Suomessa Venäjänvallan aikana." In *Suomen maanmittauksen historia II: Venäjänvallan ja itsenäisyyden aika.* Porvoo: Werner Södeström osakeyhtiö.
Norenzayan, Ara, and Hansen, Ian G. 2006. "Belief in Supernatural Agents in the Face of Death." *Personality and Social Psychology Bulletin* 32:174 – 187.
Norenzayan, Ara, and Shariff, Azim F. 2008. "The Origin and Evolution of Religious Prosociality." *Science* 322:58 – 62.
Norris, Pippa, and Inglehart, Ronald. 2004. *Sacred and Secular: Religion and Politics Worldwide.* New York: Cambridge University Press.
North, Douglass C. 1985. "Transaction Cost in History." *Journal of European Economic History* 14:557 – 576.
———. 1990. *Institutions, Institutional Change, and Economic Performance.* Cambridge: Cambridge University Press.
Nowak, Martin A. 2006. "Five Rules for the Evolution of Cooperation." *Science* 314:1560 – 1563.
Nowak, Martin A., Page, Karen M., and Sigmund, Karl. 2000. "Fairness Versus Reason in the Ultimatum Game." *Science* 289:1773 – 1775.
Nummenmaa, Lauri, Glerean, Enrico, Viinikainen, Mikko, Jääskeläinen, Iiro P., Hari, Riitta, and Sams, Mikko. 2012. "Emotions promote social interaction by synchronizing brain activity across individuals." *Proceedings of the National Academy of Sciences* 109:9599 – 9604.
Oda, Ryo, Naganawa, Takuya, Yamauchi, Shinsaku, Yamagata, Noriko, and Matsumoto-Oda, Akiko. 2009. "Altruists are trusted based on non-verbal cues." *Biology Letters* 5:752 – 754.
Oda, Ryo, Yamagata, Noriko, Yabiku, Yuki, and Matsumoto-Oda, Akiko. 2009. "Altruism Can Be Assessed Correctly Based on Impression." *Human Nature* 20:331 – 341.
Ohtsuki, Hisashi, Hauert, Christoph, Lieberman, Erez, and Nowak, Martin A. 2006. "A simple rule for the evolution of cooperation on graphs and social networks." *Nature* 441:502 – 505.
Ohtsuki, Hisashi, Iwasa, Yoh, and Nowak, Martin A. 2009. "Indirect reciprocity provides only a narrow margin of efficiency for costly punishment." *Nature* 457:79 – 82.
Oliver, Roland, and Atmore, Anthony. 2001. *Medieval Africa 1250 – 1800.* Cambridge: Cambridge University Press.
Olson, Mancur. [1965] 1971. *The Logic of Collective Action: Public Goods and the Theory of Groups.* Cambridge: Harvard University Press.
Orbell, John, Goldman, Marion, Mulford, Matthew, and Dawes, Robyn. 1992. "Religion, Context, and Constraint Toward Strangers." *Rationality and Society* 4:291 – 307.
Orbell, John M., van de Kragt, Alphons J. C., and Dawes, Robyn M. 1988. "Explaining Discussion-Induced Cooperation." *Journal of Personality and Social Psychology* 54:811 – 819.
Ostrom, Elinor. 1990. *Governing the Commons: The Evolution of Institutions for Collective Action.* New York: Cambridge University Press.

Padgett, Vernon R., and Jorgenson, Dale O. 1982. "Superstition and Economic Threat: Germany, 1918–1940." *Personality and Social Psychology* 8:736–741.

Panchanathan, Karthik, and Boyd, Robert. 2003. "A tale of two defectors: The importance of standing for evolution of indirect reciprocity." *Journal of Theoretical Biology* 224:115–126.

———. 2004. "Indirect reciprocity can stabilize cooperation without the second-order free rider problem." *Nature* 432:499–502.

Paulhus, Delroy L. 1984. "Two-Component Models of Socially Desirable Responding." *Journal of Personality and Social Psychology* 46:598–609.

Paxson, Nathaniel D. 2004. "The Entrepreneurial Ethic of the Sikhs: Religious Signaling and the Importance of Social Capital for Trust and Exchange." http://www.religionomics.com/old/erel/S2-Archives/REC04/Paxson%20-%20Sikhs.pdf [read 27.11.2009]. Paper presented at Third Annual Conference on Religion, Economics and Culture (Kansas City, 22.–24.10.2004).

Peck, Steven L. 2004. "Simulation as experiment: A philosophical reassessment for biological modeling." *Trends in Ecology and Evolution* 19:530–534.

Peelen, Marius V., Atkinson, Anthony P., and Vuilleumier, Patrik. 2010. "Supramodal Representations of Perceived Emotions in the Human Brain." *The Journal of Neuroscience* 30:10127–10134.

Peltonen, Matti. 1992. *Talolliset ja torpparit: vuosisadan vaihteen maatalouskysymys Suomessa.* Helsinki: Suomen historiallinen seura.

———. 2006. *Lukkari Saxbergin rikos ja herännäispappilan etiikka: Mikrohistoriallinen tutkimus 1800-luvun puolivälin Keuruulta.* Helsinki: Gaudeamus.

Pentikäinen, Samuli. 1978. "Konventikkeliplakaatin tulkinta Suomessa 1800-luvun alkupuolella." In *Aate ja yhteiskunta: Mikko Juvalle omistettuja tutkielmia*, edited by Markku Heikkilä, 188–203. Helsinki: Otava.

Penton, M. James. 1997. *Apocalypse Delayed: The Story of Jehovah's Witnesses (Second Edition).* Toronto, Buffalo & London: University of Toronto Press.

Peters, Edward, ed. 1980. *Heresy and Authority in Medieval Europe: Documents in Translation.* Philadelphia: University of Pennsylvania Press.

Petranek, Charles F. 1988. "Recruitment and Commitment." *Society* 25:48–.

Plutynski, Anya. 2001. "Modeling Evolution in Theory and Practice." *Philosophy of Science* 68:225–236.

Popper, Karl. 1963. *Conjectures and Refutations: The Growth of Scientific Knowledge.* London & New York: Routledge.

Prinz, Jesse. 2004. "Which Emotions Are Basic? " In *Emotion, Evolution, and Rationality*, edited by Dylan Evans, and Pierre Cruse, 69–87. Oxford: Oxford University Press.

Puolanne, Yrjö. 1936. "Hyppääjälahkosta Valkjärven seurakunnassa." In *Suomen kirkkohistoriallisen seuran vuosikirja 1925–1928*, 270–294. Helsinki: Suomen kirkkohistoriallinen seura.

Purzycki, Benjamin, and Sosis, Richard. 2010. "Religious Concepts as Necessary Components of the Adaptive Religious System." In *Interdisciplinary Perspectives on Philosophy: Evolution and Religion*, edited by Ulrich Frey, 37–59. Marburg: Tectum Verlag.

Pyysiäinen, Ilkka. 2003. *How Religion Works: Towards a New Cognitive Science of Religion.* Leiden & Boston: Brill.

———. 2004a. “Corrupt Doctrine and Doctrinal Revival: On the Nature and Limits of the Modes Theory.” In *Theorizing Religions Past: Archaelogy, History, and Cognition*, edited by Harvey Whitehouse, and Luther H. Martin, 173–194. Walnut Creek: Altamira.

———. 2004b. *Magic, Miracles, and Religion: A Scientist's Perspective.* Walnut Creek: Altamira Press.

———. 2005a. “Religion, *Délire* and Counterintuitiveness.” *Durkheimian Studies* 11:3–10.

———. 2005b. “Yhteisö ja agenttius Durkheimin uskontoteoriassa.” *Sosiologia* 42:298–308.

———. 2008. “Jumalan reviiri: Pyhä raja ja rituaalit.” In *Pyhä: Raja, kielto ja arvo kansanomaisessa uskonnossa – Veikko Anttosen juhlakirja*, edited by Tiina Mahlamäki, Ilkka Pyysiäinen, and Teemu Taira, 278–293. Helsinki: SKS.

———. 2009. *Supernatural agents: Why we believe in souls, gods and buddhas.* New York: Oxford University Press.

Pyysiäinen, Ilkka, ed. 2010a. *Religion, Economy, and Cooperation.* Berlin & New York: De Gruyter.

Pyysiäinen, Ilkka. 2010b. “Servants of Two Masters: Religion, Economy, and Cooperation.” In *Religion, Economy, and Cooperation*, edited by Ilkka Pyysiäinen, 1–34. Berlin & New York: De Gruyter.

Pyysiäinen, Ilkka, and Hauser, Marc. 2010. “The origins of religion: evolved adaptation or by-product?” *Trends in Cognitive Sciences* 14:104–109.

Raittila, Pekka. 1976. *Lestadiolaisuus 1860-luvulla: Leviäminen ja yhteisönmuodostus.* Helsinki: Akateeminen kustannusliike.

———. 1985a. “Aatehistoriallinen selitystapa ja lähteet (vastaus Ylikankaan kirjoitukseen).” *Teologinen aikakauskirja* 90:404.

———. 1985b. “Heikki Ylikangas, Murtuva säätyvalta (kirja-arvio).” *Teologinen aikakauskirja* 90:251–252.

Randolph-Seng, Brandon, and Nielsen, Michael E. 2007. “Honesty: One Effect of Primed Religious Representations.” *The International Journal for the Psychology of Religion* 17:303–315.

Ranta, Raimo. 1978. *Pohjanmaan maaseudun käsityöläiset vuosina 1721–1809 I: Käsityöläiseksi pääsy ja käsityöläisten lukumäärä.* Helsinki: Suomen historiallinen seura.

Ranta, Raimo, and Åström, Sven-Erik. 1980. “Ulkomaan- ja kotimaankauppa.” In *Suomen taloushistoria 1: Agraarinen Suomi*, edited by Eino Jutikkala, Yrjö Kaukiainen, and Sven-Erik Åström, 266–293. Helsinki: Tammi.

Ranta, Raimo, Åström, Sven-Erik, and Myllyntaus, Timo. 1980. “Tapulipolitiikka ja Suomi.” In *Suomen taloushistoria 1: Agraarinen Suomi*, edited by Eino Jutikkala, Yrjö Kaukiainen, and Sven-Erik Åström, 255–265. Helsinki: Tammi.

Rappaport, Roy A. 1979a. “The Obvious Aspects of Ritual.” In *Ecology, Meaning & Religion*, 173–222. Berkeley: North Atlantic Books.

———. 1979b. “Sanctity and Lies in Evolution.” In *Ecology, Meaning & Religion*, 223–246. Berkeley: North Atlantic Books.

———. 1984. *Pigs for the Ancestors: Ritual in the Ecology of a New Guinea People.* New Haven & London: Yale University Press, a new, enlarged edition edn.

———. 1999. *Ritual and religion in the making of humanity.* Cambridge: Cambridge University Press.

Remes, Viljo. 1995. *Herännäisyyden nousu ja hajoaminen.* Lapua: Herättäjä-Yhdistys.

Richerson, Peter J., and Boyd, Robert. 2005. *Not by Genes Alone: How Culture Transformed Human Evolution.* Chicago & London: The University of Chicago Press.

Richerson, Peter J., Boyd, Robert T., and Henrich, Joseph. 2003. "Cultural Evolution of Human Cooperation." In *Genetic and Cultural Evolution of Cooperation*, edited by Peter Hammerstein, 355–388. Cambridge & London: The MIT Press.

Richerson, Peter J., and Newson, Lesley. 2009. "Is Religion Adaptive? Yes, No, Neutral. But Mostly We Don't Know." In *The Believing Primate: Scientific, Philosophical, and Theological Reflections on the Origin of Religion*, edited by Jeffrey Schloss, and Michael Murray, 100–117. Oxford: Oxford University Press.

Roccas, Sonia, and Brewer, Marilynn B. 2002. "Social Identity Complexity." *Personality and Social Psychology Review* 6:88–106.

Roes, Frans L., and Raymond, Michel. 2003. "Belief in moralizing gods." *Evolution and Human Behavior* 24:126–135.

Roof, Wade Clark, and Hoge, Dean R. 1980. "Church Involvement in America: Social Factors Affecting Membership and Participation." *Review of Religious Research* 21:405–426.

Roos, J. P. 2009. "Evoluutio, talouden käsitteet ja pyyteettömyys." *Tieteessä tapahtuu* 27:53–54.

Rosendal, M. 1902. *Suomen herännäisyyden historia XIX:llä vuosisadalla (ensimmäinen osa, 1796–1835)*. Oulu: Kustannusosakeyhtiö Herättäjä.

———. 1905. *Suomen herännäisyyden historia XIX:llä vuosisadalla (toinen osa, 1836–1844)*. Oulu: Kustannusosakeyhtiö Herättäjä.

———. 1912. *Suomen herännäisyyden historia XIX:llä vuosisadalla (kolmas osa, 1845–1852)*. Oulu: Kustannusosakeyhtiö Herättäjä.

———. 1915. *Suomen herännäisyyden historia XIX:llä vuosisadalla (neljäs osa, 1853–1900)*. Oulu: Kustannusosakeyhtiö Herättäjä.

Rowe, Candy. 1999. "Receiver psychology and the evolution of multicomponent signals." *Animal Behaviour* 58:921–931.

Ruffle, Bradley J., and Sosis, Richard. 2006. "Cooperation and the in-group-out-group bias: A field test on Israeli kibbutz members and city residents." *Journal of Economic Behavior & Organization* 60:147–163.

Ruiz-Belda, María-Angeles, Fernández-Dols, José-Miguel, Carrera, Pilar, and Barchard, Kim. 2003. "Spontaneous facial expressions of happy bowlers and soccer fans." *Cognition and Emotion* 17:315–326.

Ruuth, Martti. 1942. "Matthias Akiander." In *Juhlakirja Suomen kirkkohistoriallisen seuran täyttäessä 50 vuotta*, edited by Martti Ruutu, *Suomen kirkkohistoriallisen seuran toimituksia*, vol. XLVI, 129–144. Helsinki: Suomen kirkkohistoriallinen seura.

Ruutu, Martti. 1967. "De religiösa folkrörelserna och samhället ca 1750–1850: Finland." In *Nordiska historiker mötet 1, Historiallinen arkisto*, vol. 62, 95–112. Helsinki: Suomen historiallinen seura.

Saarenheimo, Juhani. 2003. "Isojako." In *Suomen maatalouden historia – Osa 1: Perinteisen maatalouden aika, esihistoriasta 1870-luvulle*, edited by Viljo Rasila, Eino Jutikkala, and Anneli Mäkelä-Alitalo, 349–364. Helsinki: Suomalaisen kirjallisuuden seura.

Sales, Stephen M. 1972. "Economic threat as a determinant of conversion rates in authoritarian and nonauthoritarian churches." *Journal of Personality and Social Psychology* 23:420–428.

Sally, David. 1995. "Conversation and Cooperation in Social Dilemmas: A Meta-analysis of Experiments from 1958 to 1992." *Rationality and Society* 7:58–92.

Salomies, Ilmari. 1962. *Suomen kirkon historia III: Isonvihan kynnykseltä Ruotsin vallan loppuun*. Helsinki: Kustannusyhtiö Otava.

Salomäki, Hanna. 2010. *Herätysliikkeisiin sitoutuminen ja osallistuminen*. Tampere: Kirkon tutkimuskeskus.

Sanderson, Stephen K. 2008. "Adaptation, evolution, and religion." *Religion* 38:141–156.

Schachter, Stanley, and Singer, Jerome E. 1962. "Cognitive, Social, and Physiological Determinants of Emotional State." *Psychological Review* 69:379–399.

Schloss, Jeffrey, and Murray, Michael, eds. 2009. *The Believing Primate: Scientific, Philosophical, and Theological Reflections on the Origin of Religion*. Oxford: Oxford University Press.

Schloss, Jeffrey P. 2008. "He Who Laughs Best: Involuntary Religious Affect as a Solution to Recursive Cooperative Defection." In *The Evolution of Religion: Studies, Theories, and Critiques*, edited by Joseph Bulbulia, Richard Sosis, Erica Harris, Russell Genet, Cheryl Genet, and Karen Wyman, 197–210. Santa Margarita: Collins Foundation Press.

Schloss, Jeffrey P., and Murray, Michael J. 2011. "Evolutionary accounts of belief in supernatural punishment: a critical review." *Religion, Brain & Behavior* 1:46–99.

Schmaus, Warren. 2010. "Durkheim and Psychology." In *Religion, Economy, and Cooperation*, edited by Ilkka Pyysiäinen, 99–126. Berlin & New York: De Gruyter.

Schybergson, Per. 1980. "Teollisuus ja käsityö." In *Suomen taloushistoria 1: Agraarinen Suomi*, edited by Eino Jutikkala, Yrjö Kaukiainen, and Sven-Erik Åström, 408–435. Helsinki: Tammi.

Seidl, David, and Becker, Kai Helge. 2006. "Organizations as Distinction Generating and Processing Systems: Niklas Luhmann's Contribution to Organization Studies." *Organization* 13:9–35.

Seppo, Juha. 1987. "Seurakuntaelämä ja uskonnolliset liikkeet Etelä-Pohjanmaalla 1809–1917." In *Etelä-Pohjanmaan historia 6: Autonomian kausi 1809–1917*. Vaasa: Etelä-Pohjanmaan maakuntaliitto.

Shariff, Azim F., and Norenzayan, Ara. 2007. "God Is Watching You: Priming God Concepts Increases Prosocial Behavior in an Anonymous Economic Game." *Psychological Science* 18:803–809.

———. 2011. "Mean Gods Make Good People: Different Views of God Predict Cheating Behavior." *The International Journal for the Psychology of Religion* 21:85–96.

Shariff, Azim F., Norenzayan, Ara, and Henrich, Joseph. 2009. "The Birth of High Gods: How the Cultural Evolution of Supernatural Policing Agents Influenced the Emergence of Complex, Cooperative Human Societies, Paving the Way for Civilization." In *Evolution, Culture and the Human Mind*, edited by Mark Schaller, Ara Norenzayan, Steven J. Heine, Toshio Yamagishi, and Tatsuya Kameda, 119–136. New York & London: Psychology Press.

Sigmund, Karl. 2007. "Punish or perish? Retaliation and collaboration among humans." *Trends in Ecology and Evolution* 22:593–600.

Siltala, Juha. 1992. *Suomalainen ahdistus: Huoli sielun pelastumisesta*. Helsinki: Kustannusosakeyhtiö Otava.

Simmonds, Robert B. 1977. "Conversion or Addiction: Consequences of Joining a Jesus Movement Group." *The American Behavioral Scientist* 20:909–924.

Smilde, David, and May, Matthew. 2010. "The Emerging Strong Program in the Sociology of Religion." http://www.ssrc.org/blogs/immanent_frame/wp-content/uploads/2010/02/Emerging-Strong-Program-TIF.pdf [read 11. 2. 2010].

Smoak, Gregory E. 2006. *Ghost Dances and Identity: Prophetic Religion and American Indian Ethnogenesis in the Nineteenth Century.* Berkeley, Los Angeles & London: University of California Press.

Snarey, John. 1996. "The Natural Environment's Impact upon Religious Ethics: A Cross-Cultural Study." *Journal for the Scientific Study of Religion* 35:85–96.

Sober, Elliott, and Wilson, David Sloan. 1998. *Unto Others: The Evolution and Psychology of Unselfish Behavior.* Cambridge & London: Harvard University Press.

Soininen, Arvo. 1974. *Vanha maataloutemme: Maatalous ja maatalousväestö Suomessa perinnäisen maatalouden loppukaudella 1720-luvulta 1870-luvulle, Historiallisia tutkimuksia*, vol. 96. Helsinki: Suomen historiallinen seura.

———. 1980. "Maatalous." In *Suomen taloushistoria 1: Agraarinen Suomi*, edited by Eino Jutikkala, Yrjö Kaukiainen, and Sven-Erik Åström, 386–407. Helsinki: Tammi.

Soler, Montserrat. 2008. "Commitment Costs and Cooperation: Evidence from Candomble, an Afro Brazilian Religion." In *The Evolution of Religion: Studies, Theories, and Critiques*, edited by Joseph Bulbulia, Richard Sosis, Erica Harris, Russell Genet, Cheryl Genet, and Karen Wyman, 167–174. Santa Margarita: Collins Foundation Press.

Soltis, Joseph, Boyd, Robert, and Richerson, Peter J. 1995. "Can Group-functional Behaviors Evolve by Cultural Group Selection? An Empirical Test." *Current Anthropology* 36:473–494.

Sosis, Richard. 2000. "Religion and intra-group cooperation: preliminary results of a comparative analysis of utopian communities." *Cross-Cultural Research* 34:70–87.

———. 2003a. "Book Review of Darwin's Cathedral: Evolution, Religion, and the Nature of Society by David Sloan Wilson." *Evolution and Human Behavior* 24:137–143.

———. 2003b. "Why Aren't We All Hutterites? Costly Signaling Theory and Religious Behaviour." *Human Nature* 14:91–127.

———. 2005. "Does Religion Promote Trust? The Role of Signaling, Reputation, and Punishment." *Interdisciplinary Journal of Research on Religion* 1:1–30.

———. 2006. "Religious Behaviors, Badges, and Bans: Signaling Theory and the Evolution of Religion." In *Where God and Science Meet – How Brain and Evolutionary Studies Alter Our Understanding of Religion, Volume 1: Evolution, Genes, and the Religious Brain*, edited by Patrick McNamara, 61–86. Westport: Praeger Publishers.

———. 2007. "Psalms for Safety: Magico-Religious Responses to Threats of Terror." *Current Anthropology* 48:903–911.

———. 2009. "The Adaptationist-Byproduct Debate on the Evolution of Religion: Five Misunderstandings of the Adaptationist Program." *Journal of Cognition and Culture* 9:339–356.

Sosis, Richard, and Bressler, Eric. 2003. "Cooperation and commune longevity: A test of the costly signaling theory of religion." *Cross-Cultural Research* 37:211–239.

Sosis, Richard, and Bulbulia, Joseph. 2011. "The behavioral ecology of religion: the benefits and costs of one evolutionary approach." *Religion* 41:341–362.

Sosis, Richard, Kress, Howard C., and Boster, James S. 2007. "Scars for war: Evaluating alternative signaling explanations for cross-cultural variance in ritual costs." *Evolution and Human Behavior* 28:234–247.

Sosis, Richard, and Ruffle, Bradley J. 2003. "Religious Ritual and Cooperation: Testing for a Relationship on Israeli Religious and Secular Kibbutzim." *Current Anthropology* 44:713–722.

———. 2004. "Ideology, Religion, and the Evolution of Cooperation: Field Experiments on Israeli Kibbutzim." *Research in Economic Anthropology* 23:89–117.

Sosis, Richard, and Swartwout, Paul. 2008. "Demonstrating group selection: A comment on Janet Landa's 'The bioeconomics of homogenous middleman groups as adaptive units'." *Journal of Bioeconomics* 10:297–301.

Spence, Michael. 1973. "Job Market Signaling." *The Quarterly Journal of Economics* 87:355–374.

Sperber, Dan. 1996. *Explaining Culture: A Naturalistic Approach.* Oxford: Blackwell.

Stark, Rodney. 1996. "Religion as Context: Hellfire and Delinquency One More Time." *Sociology of Religion* 57:163–173.

———. 1997. *The Rise of Christianity: How the Obscure, Marginal Jesus Movement Became the Dominant Religious Force in the Western World in a Few Centuries.* San Francisco: HarperSanFrancisco.

Stark, Rodney, and Bainbridge, William Sims. 1980. "Networks of Faith: Interpersonal Bonds and Recruitment to Cults and Sects." *American Journal of Sociology* 85:1376–1395.

———. 1985. *The Future of Religion: Secularization, Revival and Cult Formation.* Berkeley, Los Angeles & London: University of California Press.

———. [1987] 1996. *A Theory of Religion.* New Brunswick: Rutgers University Press.

Stark, Rodney, and Iannaccone, Laurence R. 1997. "Why the Jehovah's Witnesses Grow so Rapidly: A Theoretical Application." *Journal of Contemporary Religion* 12:133–157.

Stearns, Stephen C. 2007. "Are We Stalled Part Way Through a Major Evolutionary Transition from Individual to Group?" *Evolution* 61:2275–2280.

Strassmann, Beverly I. 1996. "Menstrual hut visits by Dogon women: a hormonal test distinguishes deceit from honest signaling." *Behavioral Ecology* 7:304–315.

Strauss, Claudia, and Quinn, Naomi. 1997. *A Cognitive Theory of Cultural Meaning.* Cambridge: Cambridge University Press.

Stroope, Samuel. 2011. "Education and religion: Individual, congregational, and cross-level interaction effects on biblical literalism." *Social Science Research* 40:1478–1493.

Sugden, Robert. 2000. "Credible worlds: the status of theoretical models in economics." *Journal of Economic Methodology* 7:1–31.

Sulkunen, Irma. 1983. "Väckelserörelserna som ett förskede i organiseringens historia." *Historisk tidskrift för Finland* 1–14.

———. 1999. *Liisa Eerikintytär ja hurmosliikkeet 1700–1800-luvulla.* Helsinki: Gaudeamus.

Sun, Ron, ed. 2006. *Cognition and Multi-Agent Interaction: From Cognitive Modeling to Social Simulation.* Cambridge: Cambridge University Press.

Sun, Ron, and Naveh, Isaac. 2007. "Social institution, cognition, and survival: A cognitive-social simulation." *Mind and Society* 6:115–142.

Sundberg, Kerstin. 1993. *Resurser och sociala relationer – Studier av ett lokalsamhälle i förändring 1600–1800: Österhaninge och Västerhaninge socknar.* Lund: Lund University Press.

Suolahti, Gunnar. 1919. *Suomen papisto 1600- ja 1700-luvuilla.* Porvoo: Werner Söderström osakeyhtiö.

Suolinna, Kirsti. 1975a. "Herätysliikkeet sosiaalisina liikkeinä." In *Uskonnollinen liike: suomalaisia tapaustutkimuksia*, edited by Juha Pentikäinen, 28–56. Helsinki: Suomalaisen kirjallisuuden seura.

———. 1975b. *Uskonnollisten liikkeitten asema sosiaalisessa muutoksessa.* Helsinki: Helsingin yliopiston sosiologian laitos.

———. 1975c. "Yhteiskuntarakenne ja uskonnollinen liike." In *Uskonnollinen liike: suomalaisia tapaustutkimuksia*, edited by Juha Pentikäinen, 57–76. Helsinki: Suomalaisen kirjallisuuden seura.

———. 1977. "Lestadiolaisuus ja agraarin väestön puolustusmekanismi." In *Maailmankuvan muutos tutkimuskohteena: näkökulmia teollistumisajan Suomeen*, edited by Matti Kuusi, Risto Alapuro, and Matti Klinge, 112–120. Helsinki: Kustannusosakeyhtiö Otava.

———. 1982. "Pohjoismaiden herätysliikkeet kansanliikkeinä." In *Kirkkososiologia*, edited by Irja Askola, Tapio Lampinen, and Pentti Lempiäinen, 90–108. Helsinki: Kirjapaja.

Tajfel, Henri. 1970. "Experiments in Intergroup Discrimination." *Scientific American* 223:96–102.

Talonen, Jouko. 1993. *Esikoislestadiolaisuus ja suomalainen yhteiskunta 1900–1944*. Helsinki: Suomen kirkkohistoriallinen seura.

Tan, Jonathan H. W., and Vogel, Claudia. 2008. "Religion and trust: An experimental study." *Journal of Economic Psychology* 29:832–848.

Tautz, Jürgen. 2009. *The Buzz About Bees: Biology of a Superorganism*. Berlin: Springer.

Taylor, Michael. 1982. *Community, Anarchy & Liberty*. Cambridge: Cambridge University Press.

Teller, Paul. 2001. "Twilight of the Perfect Model Model." *Erkenntnis* 55:393–415.

Thornton, Russell. 1981. "Demographic Antecedents of a Revitalization Movement: Population Change, Population Size, and the 1890 Ghost Dance." *American Sociological Review* 46:88–96.

Toivanen, Hannes. 1997. *Yksilöllisen ja erillisen synty: kollektiivisen maanhallinnan kriisi Merikarvialla vuosina 1750–1800*. Helsinki: Helsingin yliopisto, Historian laitos. Unpublished master's thesis.

Tooby, John, and Cosmides, Leda. 1992. "The Psychological Foundations of Culture." In *The Adapted Mind: Evolutionary Psychology and the Generation of Culture*, edited by Jerome Barkow, Leda Cosmides, and John Tooby, 19–136. New York: Oxford University Press.

———. 2008. "The evolutionary psychology of the emotions and their relationship to internal regulatory variables." In *Handbook of Emotions*, edited by M. Lewis, J. M. Haviland-Jones, and L. F. Barrett, 114–137. New York: Guilford Press.

Torvi. 2008. "Verkosto." http://rauhanyhdistys.blogspot.com/2008/07/verkosto.html [read 16. 8. 2011].

Trimble, Douglas E. 1997. "The Religious Orientation Scale: Review and Meta-Analysis of Social Desirability Effects." *Educational and Psychological Measurement* 57:970–986.

Trivers, Robert L. 1971. "The Evolution of Reciprocal Altruism." *The Quarterly Review of Biology* 46:35–57.

———. 1998. "As they would to you (review of "Unto Others" by Sober and Wilson)." *Skeptic* 6:81–83.

Troberg, Martti. 1995. *"Ompa meilläkin vaunut, joilla taivasehen vierään!": Vasulaiset – kansanliikkeen rakenne ja kehitys Kuortaneella v. 1799–1830*. Helsinki: Helsingin yliopisto, Historian laitos. Unpublished master's thesis.

Turchin, Peter. 2003. *Complex Population Dynamics: A Theoretical/Empirical Synthesis*. Princeton: Princeton University Press.

———. 2009. "A theory for formation of large empires." *Journal of Global History* 4:191–217.

Turchin, Peter, and Nefedov, Sergey A. 2009. *Secular Cycles*. Princeton & Oxford: Princeton University Press.

Turiel, Elliot. 1998. "The Development of Morality." In *Handbook of Child Psychology (Volume 3: Social, emotional, and personality development)*, edited by William Damon, and Nancy Eisenberg, 863–932. New York: Wiley.

Vainio-Korhonen, Kirsi. 1998. *Käsin tehty – miehelle ammatti, naiselle ansioiden lähde: Käsityötuotannon rakenteet ja strategiat esiteollisessa Turussa Ruotsin ajan lopulla.* Helsinki: Suomen Historiallinen Seura.

———. 2002. *Ruokaa, vaatteita, hoivaa: Naiset ja yrittäjyys paikallisena ja yleisenä ilmiönä 1700-luvulta nykypäivään.* Helsinki: Suomalaisen kirjallisuuden seura.

Veblen, Thorstein. [1899] 2004. *The Theory of the Leisure Class.* Whitefish: Kessinger Publishing.

Virrankoski, Pentti, and Åström, Sven-Erik. 1980. "Teollisuus ja käsityö." In *Suomen taloushistoria 1: Agraarinen Suomi*, edited by Eino Jutikkala, Yrjö Kaukiainen, and Sven-Erik Åström, 240–254. Helsinki: Tammi.

Visala, Aku. 2009. *Religion Explained? A Philosophical Appraisal of the Cognitive Science of Religion.* Helsinki: University of Helsinki (unpublished academic dissertation).

Vrij, Aldert, Edward, Katherine, Roberts, Kim P., and Bull, Ray. 2000. "Detecting Deceit via Analysis of Verbal and Nonverbal Behavior." *Journal of Nonverbal Behavior* 24:239–263.

Vytal, Katherine, and Hamann, Stephan. 2010. "Neuroimaging Support for Discrete Neural Correlates of Basic Emotions: A Voxel-based Meta-analysis." *Journal of Cognitive Neuroscience* 22:2864–2885.

Wallace, Anthony F. C. 1956. "Revitalization Movements." *American Anthropologist* 58:264–281.

Weber, Max. 1930. *The Protestant Ethic and the Spirit of Capitalism.* London: George Allen & Unwin Ltd.

West, Stuart, el Mouden, Claire, and Gardner, Andy. 2011. "Sixteen common misconceptions about the evolution of cooperation in humans." *Evolution and Human Behavior* 32:231–262.

Whitehouse, Harvey. 2000. *Aguments and Icons: Divergent Modes of Religiosity.* Oxford: Oxford University Press.

Whitehouse, Harvey, and Martin, Luther H., eds. 2004. *Theorizing Religions Past: Archaelogy, History, and Cognition.* Walnut Creek: Altamira.

Wiessner, Polly. 2009. "Experimental Games and Games of Life among the Ju/'hoan Bushmen." *Current Anthropology* 50:133–138.

Wildman, Wesley J., and Sosis, Richard. 2011. "Stability of Groups with Costly Beliefs and Practices." *Journal of Artificial Societies and Social Simulation* 14. http://jasss.soc.surrey.ac.uk/14/3/6.html [read 17.8.2011].

Wilensky, Uri. 1999. "NetLogo." http://ccl.northwestern.edu/netlogo/. Center for Connected Learning and Computer-Based Modeling, Northwestern University, Evanston, IL.

Wilska, Terhi-Anna, ed. 2006. *Uskon asia: Nuorisobarometri 2006.* Helsinki: Nuorisoasiain neuvottelukunta, Nuorisotutkimusverkosto & Opetusministeriö.

Wilson, Bryan R. 1990. *The Social Dimensions of Sectarianism: Sects and New Religious Movements in Contemporary Society.* Oxford: Clarendon Press.

Wilson, David Sloan. 2002. *Darwin's Cathedral: Evolution, Religion, and the Nature of Society.* Chicago & London: The University of Chicago Press.

———. 2005. "Testing Major Evolutionary Hypotheses about Religion with a Random Sample." *Human Nature* 16:419–446.

Wilson, Edward O. 1975. *Sociobiology: The New Synthesis*. Cambridge & London: The Belknap Press of Harvard University Press.

Wilson, Timothy D., Gilbert, Daniel T., and Centerbar, David B. 2002. "Making Sense: The Causes of Emotional Evanescence." In *Economics and Psychology*, edited by J. Carrillo, and I. Brocas, 209–233. New York: Oxford University Press.

Wilson-Mendenhall, Christine D., Barrett, Lisa Feldman, Simmons, W. Kyle, and Barsalou, Lawrence W. 2011. "Grounding emotion in situated conceptualization." *Neuropsychologia* 49:1105–1127.

Wiltermuth, Scott S., and Heath, Chip. 2009. "Synchrony and Cooperation." *Psychological Science* 20:1–5.

Winsberg, Eric. 2003. "Simulated Experiments: Methodology for a Virtual World." *Philosophy of Science* 70:105–125.

Winterhalder, Bruce, and Smith, Eric Alden. 2000. "Analyzing Adaptive Strategies: Human Behavioral Ecology at Twenty-Five." *Evolutionary Anthropology* 9:51–72.

Wirilander, Kaarlo. 1974. *Herrasväkeä: Suomen säätyläistö 1721–1870*. Helsinki: Suomen historiallinen seura.

Yamagishi, Toshio, Jin, Nobuhito, and Kiyonari, Toko. 1999. "Bounded generalized reciprocity: Ingroup favoritism and ingroup boasting." *Advances in Group Processes* 16:161–197.

Yamagishi, Toshio, Tanidab, Shigehito, Mashimab, Rie, Shimomab, Eri, and Kanazawac, Satoshi. 2003. "You can judge a book by its cover: Evidence that cheaters may look different from cooperators." *Evolution and Human Behavior* 24:290–301.

Ylikangas, Heikki. 1979. *Körttiläiset tuomiolla: Massaoikeudenkäynnit heränneitä vastaan Etelä-Pohjanmaalla 1830- ja 1840-lukujen taitteessa*. Helsinki: Kustannusosakeyhtiö Otava.

———. 1984. *Murtuva säätyvalta*. Porvoo, Helsinki & Juva: Werner Söderström Osakeyhtiö.

———. 1985. "Aatehistoriallisen selitystavan riittävyys (vastaus Raittilan kirja-arvosteluun)." *Teologinen aikakauskirja* 90:294.

———. 1989. *Tervasta teollisuuteen: Ylihärmän kehitys varhaisista ajoista nykypäiviin*. Ylihärmä: Ylihärmän kunta.

———. 1990a. "Herätysliikkeiden pohjasta ja perustasta." In *Kirkko ja politiikka: Juhlakirja professori Eino Murtorinteen täyttäessä 60 vuotta 25.11.1990*, edited by Hannu Mustakallio, 320–343. Helsinki: Suomen kirkkohistoriallinen seura.

———. 1990b. *Mennyt meissä: Suomalaisen kansanvallan historiallinen analyysi*. Helsinki: WSOY.

———. 1991. "Oppihistorian taloudellis-sosiaalinen perusta." In *Maaseutuväkeä, hengen ja ruumiin kulttuuria: kokoelma artikkeleita Antero Heikkisen tutkimusaloilta*, edited by Tapio Hämynen, and Antti Laine, *Historian tutkimuksia*, vol. 9, 43–51. Joensuu: Joensuun yliopisto.

———. 2002. "From witchcraft to revivalist movements: On the support for revivalist meetings and their background in Ostrobothnia in the 1830s and 1840s." In *Manslaughter, Fornication and Sectarianism*, edited by Anu Koskivirta, and Sari Forsström, 9–46. Helsinki: Academia Scientiarum Fennica.

Ylikoski, Petri. 2004. "Rationaalinen valinta ja uskonnonsosiologia: Kommentti Petri Luomasen artikkeliin." *Teologinen aikakauskirja* 109:545–548.

———. 2011. "Solidaarisuus ja itsekkyyden normi." In *Solidaarisuus*, edited by Arto Laitinen, and Anne Birgitta Pessi, 50–60. Helsinki: Gaudeamus.

Zahavi, Amotz. 1975. "Mate Selection – A Selection for a Handicap." *Journal of Theoretical Biology* 53:205–214.

Åström, Sven-Erik, Mauranen, Tapani, Schybergson, Per, and Kaukiainen, Yrjö. 1980. "Kauppa ja liikenne." In *Suomen taloushistoria 1: Agraarinen Suomi*, edited by Eino Jutikkala, Yrjö Kaukiainen, and Sven-Erik Åström, 436–483. Helsinki: Tammi.

Index